Versatile rural transportation in north-east Thailand; the vehicle's engine also doubles as a water pump and an electricity generator.

Routledge Introductions to Development

Series Editors:
John Bale and David Drakakis-Smith

Population Movements and the Third World

In the same series

Rajesh Chandra
Industrialization and Development in the Third World

John Cole
Development and Underdevelopment
A profile of the Third World

David Drakakis-Smith
The Third World City

Allan and Anne Findlay
Population and Development in the Third World

Avijit Gupta
Ecology and Development in the Third World

John Lea
Tourism and Development in the Third World

John Soussan
Primary Resources and Energy in the Third World

Chris Dixon
Rural Development in the Third World

Alan Gilbert
Latin America

Janet Henshall Momsen
Women and Development in the Third World

David Drakakis-Smith
Pacific Asia

Mike Parnwell

Population Movements and the Third World

London and New York

For Jip, Mark and Chris

First published 1993
by Routledge
11 New Fetter Lane, London EC4P 4EE

Simultaneously published in the USA and Canada
by Routledge
29 West 35th Street, New York, NY 10001

Typeset by J&L Composition Ltd, Filey, North Yorkshire
Printed and bound in Great Britain by
Biddles Ltd, Guildford and King's Lynn

British Library Cataloguing-in-Publication Data
A catalogue record for this book is available from the British Library.

ISBN 0–415–06953–X

Library of Congress Cataloging in Publication Data
Parnwell, Mike.
Population movements and the Third World / Mike Parnwell.
p. cm.—(Routledge introductions to development)
Includes bibliographical references and index.
ISBN 0–415–06953–X: $12.95
1. Migration, Internal—Developing countries. 2. Developing countries—Emigration and immigration. I. Title. II. Series.
HB2160.P37 1993
304.8′09172′4—dc20 92–1368
CI

Contents

Plates

Figures

Tables

Preface

Population movement, development and the Third World are among the broadest subjects one could imagine to have to write about. The 'Third World' exists mainly in people's minds, and is difficult to delineate or characterize in reality; 'development' means different things to different people; and 'population movement' covers a wide and complex range of forms of mobility which often have little in common. And yet population movements can also be argued to be of central importance to the development process in the Third World. They often provide the 'litmus test' of unsuccessful development, and at the same time hold the potential to counteract some of the iniquities which are often associated with the process of development in Third World countries.

The aim of this book is introduce the reader to a variety of aspects of population movement in and from Third World countries, and to link the process of movement to the features of development in these countries. Because of its importance in a great many Third World countries, the bulk of the volume is concerned with the process of migration in its various guises. Quite severe constraints of space have meant that depth and detail have unavoidably been compromised by the desire to cover as wide a range of forms of movement as possible.

Readers should be aware of the limitations of a book which purports to describe and explain such a broad and diverse set of phenomena. One can just as easily find examples to refute as to support a great many of the statements concerning population movements which are made in this volume. Whilst therefore cautioning against the use of stereotypes and generalizations to view Third World population movements, the author is very much aware that he has been guilty of conveying a number of such images of the Third World in general.

The author would like to thank a number of people who have assisted in various ways in the task of assembling this volume. Particularly warm thanks go to my wife, Jip, and children, Mark and Christopher, for uncomplainingly allowing me the time to write this book and for tolerating the frustrations which occasionally accompanied the task. Thanks also to Pauline Kh'ng for fishing out some useful references, to Terry King and David Drakakis-Smith for their support, and to Paul Lightfoot for fuelling my original interest in migration. Sincere gratitude also goes to the villagers of four communities in Roi-et province, North-East Thailand, where the author undertook field-research on return-migration, for their kindness and generosity, and to Dusit Srisaphum for the guidance and insight he provided.

Mike Parnwell
March 1992

1 Introduction

Between a rock and a hard place?

'Between a rock and a hard place' is one of several aphorisms which was used to characterize the plight of the half-million or so Asian and North African refugees who fled to Jordan from Kuwait and Iraq with the onset of the Gulf Crisis in late 1990. It aptly described the paradoxical situation whereby huge numbers of people seeking refuge from potential conflict found themselves faced with desperate conditions in hastily constructed refugee camps until the international relief operation creaked into action. Whilst the world's attention and sympathy was momentarily drawn to their plight, few people may be aware that these economic migrants-turned-refugees have continued to suffer the after-shocks of the Gulf War long after the cessation of hostilities. Having returned to their countries of origin with next to nothing, their lucrative work contracts terminated, most face very bleak prospects of finding gainful employment back home. Many are saddled with huge debts incurred in financing their sojourn in the Middle East. Somewhat ironically, the reconstruction programme following the devastation of the Gulf War offers their best chance of recouping their losses.

A key aspect of many forms of population movement in the Third World is that they offer the tantalizing chance of better prospects elsewhere, but that these very often fail to materialize in reality. Occasional references appear in the migration literature to the perception of a 'promised land' in the mind's eye of emigrants to the industrialized West, and the expectation amongst rural migrants that

Plates 1.1 and 1.2 Image and reality: an illustration of the two extremes of life in the Third World city. The modern, high-rise environment may attract migrants to urban areas but many struggle to obtain an acceptable standard of living

the streets of their capital cities, metaphorically, will be 'paved with gold'. The reality is often quite different. Immigrants may be seen and treated as second class citizens by the host society, and may have great difficulty in obtaining well-paid employment and in adjusting socially and psychologically to their move. The ubiquitous shanty towns, and the frequent images of people scavenging and otherwise scraping a living in the major urban centres of the Third World are further testament to the chasm which sometimes separates expectation from reality for many city-bound migrants from the countryside.

Whatever its outcome, however, it is hard to overstate the importance of mobility to the peoples of the Third World. The ability to move from one place to another, be it to escape the effects of environmental disasters or to exploit opportunities which may be available elsewhere, represents an essential means of dealing with the problems which beset many who live in the world's poorer countries. Migration from village to city may constitute a 'pressure valve' whereby people may escape the drudgery and uncertainty of rural life. Flight across international frontiers may, as a last resort, represent the only means by which the persecuted, the frightened and the neglected can escape warfare or the brutality of the tyrannical regimes which litter the developing world. The periodic movement of farmers and pastoralists may enable them to overcome environmental constraints on their livelihood.

On a global scale, the movement and interaction of people has, historically, been an important factor in world civilization, in the enrichment of cultures and in the spread of technology; migration representing 'an integral and vital part of human development' (Bilsborrow, *et al.*, 1984: 2). Migration has also played an important role in the processes of industrialization and urbanization, initially in Western Europe and more recently in many other parts of the globe. Conversely, migration has also provided the cornerstone of some of the less positive chapters in human history, such as the slave trade between Africa and the Americas, and forced labour in support of the economic exploitation of colonial territories.

Whatever the historical and contemporary importance of population mobility, we must not lose sight of the fact that not everyone is 'mobile' to the same extent. The wealthy and well-connected may be in a position to use their movement to another place as a means of enhancing their present circumstances and status. Others have little choice, and must make do as best they can upon arrival in a new location. Others still may not even have the option of movement, perhaps because they cannot afford to leave, or they know nothing about places and opportunities outside the confines of their home area, or possibly because of social and cultural ties which bind them to their place of origin. This latter group of people – those who lack mobility, for whatever reason – must thus endure conditions in their home areas as best they can. Many, of course, have no desire to leave in the first place.

If we consider that there are huge numbers of people in Third World countries who have been displaced from their home areas, either voluntarily or involuntarily, and that there are maybe several times as many who, given the choice, might prefer (particularly on economic grounds) to be somewhere else, we must ask the question 'why?'. There is no simple answer, but it is reasonable to suggest that the contemporary scale of population movements in the Third World, both actual and latent, may be attributed in part to one or more of the following factors:

1 The failure of the process of economic growth to bring about an even pattern of development in which all areas, sectors and groups of people have shared to a more or less equal extent. The fact that the predominant direction of movement tends to be from economically depressed areas where opportunities for advancement (or even survival) are very limited, to economically dynamic locations where opportunities are perceived to be plentiful, suggests a close association between the unevenness of the development process and the incidence of population movements (especially migration). In this sense, we might consider the incidence of migration to represent the 'litmus test' of unsuccessful development.

   As we shall see in Chapter 4, some people may be displaced from their home areas by sheer poverty and lack of opportunity, whereas others may choose to leave because they perceive that their aspirations are more likely to be satisfied somewhere other than their local area. The word 'perceive' is used advisedly here: quite often media such as television, radio, newspapers and advertising are responsible for raising people's awareness of life outside the narrow confines of their home area to the extent that it heightens their level of dissatisfaction with their present conditions and may make them more prone to migration. The image created by the media may present a biased or distorted view by building up certain positive characteristics and playing down others.

2 The penetration of the market economy has also strongly underpinned the movement of people within Third World countries. Cash

has replaced barter as the main medium of exchange in most parts of the Third World, and now perhaps the majority of households have need of cash to purchase goods and services which previously they may have provided for themselves. Out-migration to seek paid employment may represent the best (and in some instances the only) means of obtaining an income with which to satisfy these growing cash needs and people's rising aspirations with regard to preferred standards of living.

The operation of a land market, and the growing tendency for land to be seen as a commodity rather than a communal resource, has also provided the opportunity for wealthy and powerful people to accumulate vast land-holdings, and for others to dispose of their land for reasons of debt, illness or labour shortages. The associated phenomenon of landlessness, and the weak position of the land-poor *vis-à-vis* the land-rich, has also provided a strong source of momentum for the movement of people away from their home communities.

3 The processes of modernization and social change have also been important in underpinning the growth of population movements in many Third World countries, although in this case it is rather difficult to isolate cause from effect. It is really only over the last two decades or so that a number of traditional constraints on mobility, particularly those restricting the migration of young women, have started to break down with the effect that, as we shall see in Chapters 4 and 5, migration flows in several countries now consist predominantly of female migrants. It might be argued, however, that migration has itself been partly responsible for breaking down the barriers to female migration (such as culturally prescribed gender roles, or religious constraints on the participation of women in the workforce), as it has helped to raise levels of awareness about opportunities for employment elsewhere, and may also have engendered a greater level of individualism which in turn may have weakened the strength of community or parental control over the actions and decisions of young women. In several of the rapidly industrializing countries of Asia and Latin America, the rising range and number of factory jobs suitable for women may also have stimulated a growth in the incidence of female migration. Where this has occurred without a parallel process of social change, rising levels of friction and social tension in source communities may have resulted.

In many countries, education has proven a powerful instrument of social change whilst, at the same time, having the effect of raising people's qualifications and expectations beyond the capacity of their home areas to accommodate. Out-migration has very often been the inevitable result.

4 The rapid pace of population growth in many Third World countries provides further momentum to the rate of population movement, although in the majority of cases the impact is largely indirect. Thus out-migration from rural areas has often resulted from a failure or inability to accommodate population growth by expanding the cultivated area or intensifying agricultural practices. Even more alarming is the tendency for people to be displaced from the land by various forms of agricultural modernization (mechanization in particular) or through the encroachment of other forms of economic activity (commercial logging has had a severe effect on shifting cultivators in many parts of South-East Asia, for instance), when demographic conditions possibly determine that more people should be being absorbed by agriculture. The limited development and diversification of the non-farm economy may also determine that there are few employment opportunities locally which might absorb a steadily growing workforce, providing further incentive for people to move to seek their livelihood elsewhere.

5 We also find that population growth may have pushed land-hungry people into fairly marginal ecological zones such as infertile uplands, riverine areas which may be prone to flooding, or arid zones which may face prolonged periods of drought. Here it is much more difficult for people to eke out a living, and thus people may be forced into quite frequent movement simply to safeguard their economic and nutritional survival. Not infrequently, these marginal ecological zones are much more prone to natural disasters which may also result in the periodic displacement of substantial numbers of people (contemporary problems in the Horn of Africa and Bangladesh are an obvious illustration of this phenomenon).

6 People may also have to leave their homes to make way for infrastructural projects such as reservoirs, ports, roads, air terminals, and so on. In such cases the interests of the communities which are affected by these projects are usually seen as inferior to the broader national or strategic interests which may be associated with the projects' construction. None the less, the government or construction agency usually takes some degree of responsibility for the resettlement of the displaced people.

7 Finally, people throughout the Third World have been displaced from their home areas by various forms of persecution and strife. This may take the form of warfare between neighbouring countries or a civil war between rival factions; inter-ethnic or inter-religious conflicts which, in some cases may themselves have resulted from the historical movement of one group of people into the territory of another; or local disputes between individuals, families or small groups of people with others within their home communities. The nature of the political regime which governs the country may also make the position of certain groups of people in their home societies untenable. Whatever the precise reasons for the conflict, the movement of sometimes quite large numbers of people to a new location may be the only viable alternative to discomfort, danger or even death.

Whilst the movement of population may thus been seen as an essential 'release valve' for many of the Third World's contemporary problems, developmental or otherwise, it does not inevitably follow that the act of movement will lead to an improvement in the mover's predicament or prospects. Thus, many who use migration as a means of escaping poverty may find that their movement brings only a change in location, not circumstance. This is especially true of people who, although their movement may be definitionally 'voluntary', are in reality faced with no alternative since remaining *in situ* is not a realistic option. These people may have little if anything to offer and, in reality, little to gain from their movement. Even those whose departure may be entirely voluntary may fare little better from their move to a new location. This may be in part because, due to the distorted image of other places which is created by the media or which is conveyed by returning migrants, their perception of life and opportunities in the chosen destination may be quite widely divorced from reality. Thus the expectations which may have underpinned the decision to move may not be fully realized.

Refugees and other displaced peoples may similarly substitute one set of difficulties for another. Although they may have succeeded in attaining refuge or respite from whatever problems they may have been experiencing, they may encounter significant barriers to their smooth and successful settlement in another place, especially if there is resentment amongst the local population to their presence. Increasingly we find also a growing reluctance amongst governments and agencies to accept refugees, given partly the sheer scale of population displacement in many parts of the world (Cambodia and Mozambique, for example) and also the apparently growing tendency amongst 'refugees' to use the seeking of political asylum as a cover for the pursuit of better economic prospects in a second country. This very thin and difficult definitional distinction between political refugee and economic opportunist has, as we shall see in Chapter 2, controversially been used as a formidable barrier to the resettlement of Vietnamese refugees from countries of first asylum, and also a justification for their involuntary repatriation to Vietnam.

It might therefore seem somewhat illogical that people should continue to move from their home communities when, for many, their circumstances in the chosen destination may appear to be no better, and frequently much worse, than those they left behind. And yet the scale of movement would appear to be increasing. This apparent paradox may be explained by three main factors. First, a large number of movers do of course benefit substantially from their move to a new location The fact that there are potential benefits to be derived from movement thus does little to deter others from seeking to avail themselves of these benefits. Second, in reality many people have little choice but to move and thus can only hope that their movement elsewhere may lead to a change in circumstances. Unhappily, for many this does not materialize Finally, there is the question of time-scale. It is generally true, particularly in the case of rural–urban migration and emigration, that people's longer-term prospects may be rather better in their new location than in their place of origin. Thus migrants to the major urban centres of the Third World can reasonably expect their children's prospects, if not their own, for a better education, higher-paid employment and a more satisfactory living environment to be better in the city than the countryside.

If we accept that, in the short term at least, the movement of population does not automatically remedy the problems which cause people to leave their home areas, we need also to consider what effects population movement has on the broader social, structural and spatial problems which characterize a great many Third World countries. In other words, does the movement of population in general represent anything more than a basic 'survival strategy' for the people involved or does it contribute to a process whereby the longer-term prospects for development will be better than before?

It is this question which provides one of the central themes for discussion in this book. There is, of course, no simple answer. This is partly because the term 'population movements' encompasses such a wide range of forms of mobility, which occur in response to a wide and diverse range of factors, and which have such a variable and inconsistent range of effects. Thus, any attempt at generalization and simplification would be dangerously misleading. Several analysts have suggested that, on balance, the large-scale movement of population does more harm than good. In the context of the interrelationship between migration and development, John Connell has suggested that 'migration proceeds out of inequality and further establishes this inequality'. A central aim of this book is to assess the validity of such a statement.

Our main focus will be on migration within Third World countries because, in addition to being the most prevalent form of population movement, migration most clearly epitomizes the development process in the Third World which is the central theme of the *Introductions to Development* series. The volume will additionally highlight a wide range of other forms of population movement which are of contemporary significance in Third World regions. Chapter 2 will attempt to categorize the various kinds of Third World population movement in the form of a simple typology, and Chapter 3 will outline some of the more important features of the resultant types of movement.

Chapter 4 will seek to identify, at different levels of analysis, some of the reasons why migration takes place in some circumstances and not in others. Chapter 5 will consider some of the developmental effects of migration, as viewed from the perspective of the source and destination communities and the migrants themselves. Finally, Chapter 6 will assess the policy implications of contemporary patterns of migration, drawing on the experience of Third World countries in dealing with the associated problems, and will seek to suggest future directions for migration and development policy based on the evidence presented in this volume.

Key ideas

1. The movement of population in Third World countries is for many people more a 'survival strategy' than it is a mechanism for economic advancement.
2. Differences in the characteristics of population movement between the developed and developing countries are principally explained by differences in their relative development experiences.

3 The incidence of migration in the Third World can be used as the 'litmus test' for unsuccessful development.
4 How we perceive the developmental effects of population movements in the Third World depends upon our ideological viewpoint and also our perception of what 'development' should involve.
5 Whilst migration may appear rather illogical in the short term, given the problems that many migrants have to endure in their chosen destination, it is the brighter longer-term prospects which help to explain the growing incidence of migration in the Third World.

2
A typology of population movements in the Third World

The introductory chapter has given some clues as to the great complexity and variety of types of population movement which are encompassed by the title of this book. The aim of this chapter is to start to unravel this complexity by examining some of the basic elements which are involved in the movement of people from one place to another, most notably space, time and purpose. The chapter will also seek to explain some of the terms which are commonly used in the migration literature.

Any attempt at defining and simplifying such a complex web of movement types is bound to be fraught with difficulty. There is, for instance, no consensus of opinion as to the minimum and maximum spatial and temporal parameters which should be used to identify different forms of migration. How are we therefore to decide how long a person should be away from her or his home area before their absence takes on some significance for both the mover and the people who are left behind? Why should the impact of a first-time move by a young child to a neighbouring town which lasts less than a day be any less significant for the mover than the umpteenth move by an adult to find employment in another region?

In spite of problems of definition and classification, it is none the less very important to have a clear understanding of the different types of population movement in order to devise adequate and appropriate planning strategies. As we shall see, the way we define movement has a direct bearing upon the scale of movement which is identified. If the territorial unit which is used to identify the incidence of movement is too large, or the minimum time span too long, only a relatively small proportion of the total level of movement will be recorded. Where policy strategies are guided by the recorded, as opposed to the actual, level of movement they may be of little direct benefit to those whose movement does not register in mobility data. On the other hand, because of the fragmented nature of the planning system in many Third World countries, with inadequate coordination and sometimes overt competition between development agencies, planners may only be interested in certain forms of movement which may be of direct relevance to their particular area of responsibility, such as urban housing or agricultural extension. In such cases, the definitions and criteria which are employed may be guided by the kinds of movement and movers they wish to enumerate.

Some definitions

Among the terms which occur most frequently in the literature, the three Ms – mobility, movement and migration – are most central to our understanding of the processes involved, and thus we might start by attempting to define them. We often find that these terms are used interchangeably, although in reality there are some subtle differences between them.

Mobility might be considered to be the facility of being mobile, which enables some people to move from one area to another, and the absence or restricted nature of which prevents others from doing so. Mobility may thus be constrained by physical disability, but more typically it may be influenced by an individual's ability to afford the cost (psychological as well as financial) of movement, or to make the necessary arrangements which may facilitate the move. Thus we may usually find that, in the Third World context, the very poor are amongst the least mobile, particularly when it comes to availing themselves of opportunities some distance from their home areas. This may partly be because they cannot afford the cost of getting to another place and of supporting themselves upon arrival, but their mobility may also be constrained by the fact that, because of their disadvantaged situation, they may have limited access to information about potential destinations and opportunities, they may not possess the skills and qualifications needed to gain entry to the urban labour market, and they may know few people in these places who might assist in the process of movement.

Whilst there may be little more than a semantic distinction between the terms *movement* and *migration*, in this volume the latter is taken to refer to a much more restricted range of activities. Migration is generally taken to involve the permanent or quasi-permanent relocation of an individual or group of individuals from a place of origin to a place of destination. However, as we shall see shortly, there is a substantial volume of movement in the Third World which does not involve a permanent change of residence. A whole host of short-term, circular and cyclical forms of movement, which some analysts have suggested are much more prevalent than permanent migration, would not thus be included within the definition of migration which is outlined above. The distinction between 'population migration' and 'population movements' is thus a very important one, not least because of the very different characteristics and effects of these two classes of movement.

It is only during the last two decades or so that migration studies have focused on the importance of non-permanent forms of movement in the Third World, spawning a plethora of new terminologies. These are explained in diagrammatic form in Figure 2.1. Thus in order to avoid any ambiguity, the term *permanent migration* is used where the mover has no intention of returning to the place of origin, and where a lot of the migrant's energy is put into becoming established in the new location. It does not necessarily follow that all forms of contact with the migrant's natal area will be severed, however.

Whilst in the majority of cases the migration will consist of a single, unidirectional move to a chosen destination, *step migration* refers to instances where the mover arrives at a destination after a series of short-term moves to other locations, typically moving up the urban hierarchy from village to capital city. The term *emigration* is used where the mover leaves one country to settle in another.

What distinguishes other forms of migration, collectively termed *circulation*, from the above is that the migrant will, at some stage, either temporarily or permanently return to the place of origin. As we shall see shortly, the time-span between outward and return migration may range from a matter of hours to the entire working lifetime of the migrant. Shorter-term movements include *commuting*, such as occurs between home and one's place of work or education, and *oscillation*, where people move regularly to a variety of places but always return to the place of origin. 'Circulation', or *circular migration*, is generally used to refer to longer-term movements between places of origin and destination which may involve one or more cycles of outward and return movement. *Return migration* refers to the stage in the migration cycle

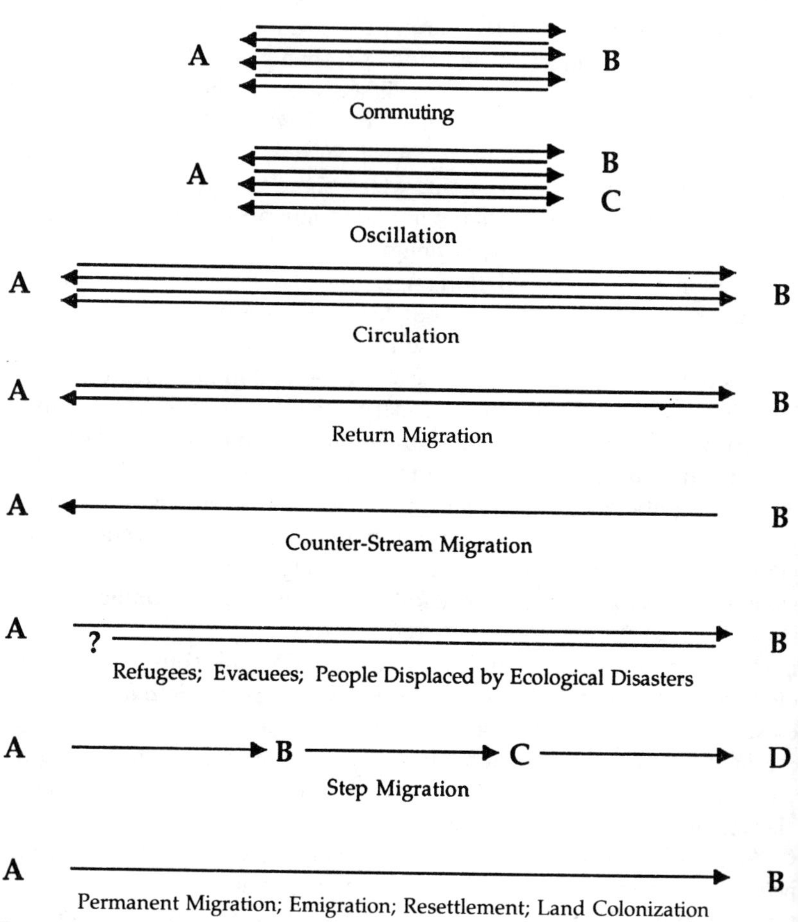

Figure 2.1 A terminological clarification of some forms of population movement

Source: Adapted from Russell King (ed.) (1986) *Return Migration and Regional Economic Problems*, p. 4, London: Croom Helm.

when the migrant leaves the destination to return to his or her natal area, and *counter-stream migration* constitutes movements in the opposite direction to the predominant streams of migration (typically from city to village or from centres of economic activity to economically depressed regions), and may consist primarily of return movements.

The factors which underlie these various forms of migration and circulation are discussed in more detail in the following sections.

A feature which all the above forms of movement have in common is that the mover generally has a free choice as to whether to move or stay. There are other forms of movement where the movement may be enforced by prevailing political, environmental or developmental circumstances. Terms which are used in association with such forms of involuntary movement include refugee, evacuee and resettlement. A *refugee* is defined by the United Nations as someone who 'owing to a well-founded fear of being persecuted for reasons of race, religion, nationality, membership of a particular social group or political opinion, is outside the country of his [or her] nationality and is unable or, owing to such fear, is unwilling to avail himself [herself] of the protection of that country.' *Evacuees* are people who have been displaced from their homes by such phenomena as natural disasters (volcanic eruptions, earthquakes, typhoons) and various infrastructural projects (such as the flooding of land to form a reservoir or the clearance of land to construct an air terminal). *Resettlement* is the process whereby such displaced people are moved to a new location and, generally, given assistance by government in order to establish themselves therein. As Figure 2.1 shows, in the majority of cases these involuntary movements will involve a permanent change in the place of residence, although for many there may be a great deal of uncertainty concerning their final destination. Many might cling to what is for most the forlorn hope of eventually returning to their original homes.

A typology of population movements

Our attempt at defining some of the more prevalent forms of population movement in the Third World has already shown that the amount of time spent away from the mover's place of origin is an important variable in helping to differentiate between various types of movement. There may also be a close correlation between the duration of the movement and the distance the mover is able to travel from the home area – this is certainly the case with short-term movements such as commuting (unless there is a very cheap and reliable transportation system).

In seeking to find some sense of order amidst the great confusion of movement types which characterize the Third World, time and space therefore provide two fundamental elements. A third important element concerns the factors which motivate people to leave their home areas. The following sections will examine these three key ingredients in the process of population movement in the Third World.

The spatial dimension

The two most significant elements of the spatial dimensions of Third World population movements are distance and direction. Movement types lie within a 'distance continuum' which ranges between very short-distance movements, which might typically take place within the mover's local area, through to very long-distance movements, which may involve migration between two countries or continents. These are only very approximate illustrations: thousands of people may make regular short-distance moves across international frontiers, particularly in remote areas where ethnic groups straddle arbitrarily drawn political boundaries.

Distance provides a useful basis for differentiating between types of movement and types of mover, because the distance over which a person travels can also be used as a proxy for other important variables. First, there is usually a close association between the distance and the cost of the movement. As cost is often important in determining whether or not people are able to migrate, and where they are able to move to, the relative distance of movements may have a filtering effect upon the kinds of people who are moving between different areas.

Second, it is generally the case that the further people move from their home areas the greater will be the contrast in social and cultural environments between which they move. There may be subtle changes in the dialects which are spoken or in the customs and norms around which social groups are organized. There may be a greater hostility to 'outsiders' where there are strongly identifiable differences between social groups, such as on the basis of ethnicity, colour or religion. This too may have the effect of filtering out people who are unwilling to venture very far from the familiarity and security of their local communities. It will also determine the extent to which people will have to change if they are to adapt to their new surroundings (the question of migrant adaptation is discussed further in Chapter 5).

The contrast in the modes of economic and social organization which exists between rural and urban areas, or in more general terms between the traditional and modern spheres, in Third World countries may also necessitate some quite fundamental forms of adaptation by people who

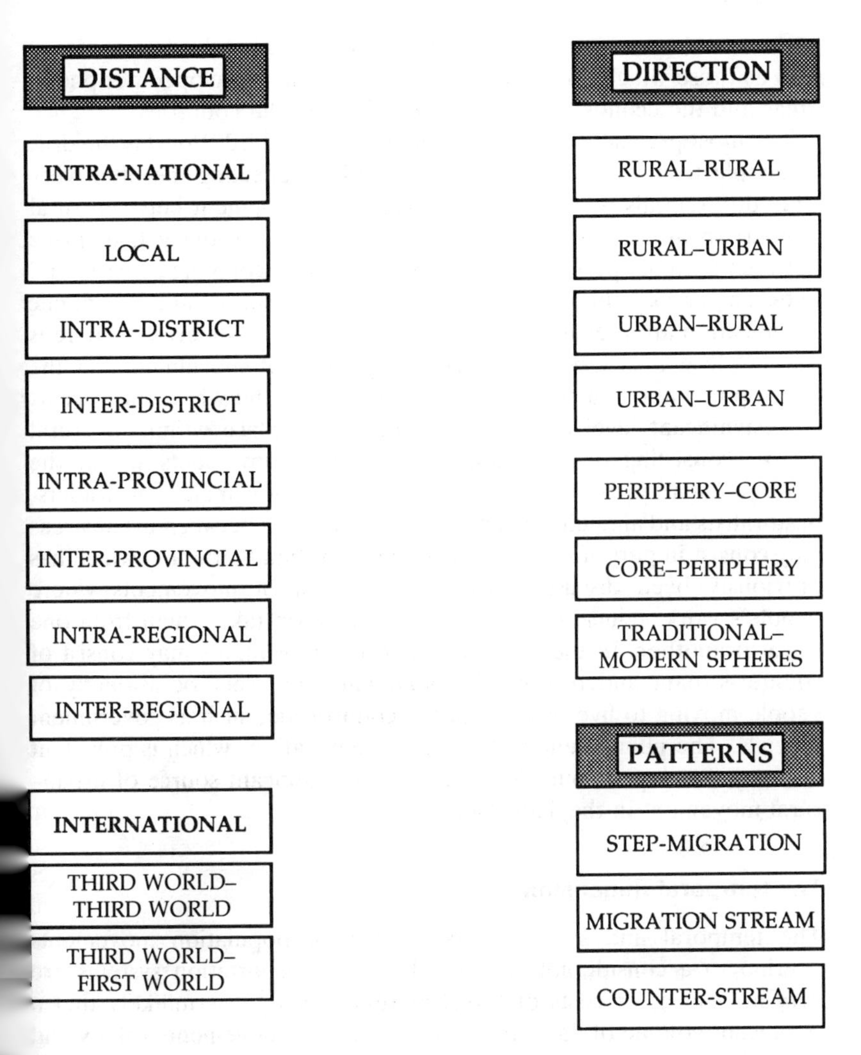

Figure 2.2 Spatial dimensions of population movement in the Third World

move between them. Because such contrasts may also be experienced by people who move only relatively short distances, the *direction* of their movement may be considered to be just as important to the classification of movement types as the distance over which they take place. Figure 2.2 shows, in a simplified way, that the main directional elements which are of importance in the Third World context are the various permutations of movements between rural and urban areas, as well as those which take place between the underdeveloped peripheral areas and the economic heartlands of Third World countries.

The most prevalent form of movement in the Third World is the drift of people from the countryside to the cities, reflecting the often very wide differentials in the level and pace of development between rural and urban areas. Such movements might simultaneously take place between national peripheries and the centres of economic activity, and between areas where traditional modes of social and economic organization are prevalent and the main centres of modernity. There is thus a close association between the pattern of movement and the developmental characteristics of the territory within which they occur.

A significant level of movement also takes place within the rural sector, consisting in the main of short-term movements within the mover's locality as well as longer-distance movements of traders, pastoralists and agricultural labourers. Movements between urban areas may consist in part of step-migration up the urban hierarchy, which has previously been discussed, or a whole host of movements where people's work, education, worship or recreation takes them from one town to another. In the main, urban–rural movements may consist of counter-stream and return migration, and only very occasionally of people moving to live or work in the countryside, such as government officials. The phenomenon of counter-urbanization, which is prevalent in many developed countries, is not yet a significant source of urban–rural movement in the Third World.

The temporal dimension

The temporal and spatial characteristics of population movements overlap to a considerable degree. Unless transportation systems are very advanced, or costs of travel extremely low, it is unlikely that a significant volume of daily and other short-term movements will extend beyond a few tens of miles. On the other hand, because of the considerable financial outlay which is required to support long-distance or international migration, it is only to be expected that in such circumstances migrants will be absent from their home area for an extended period – certainly long enough to secure an acceptable return on the initial investment. Figure 2.3 gives an impression of the link between the distance migrants travel in West Java and the period of time they are away from their home areas. Only in villages which are

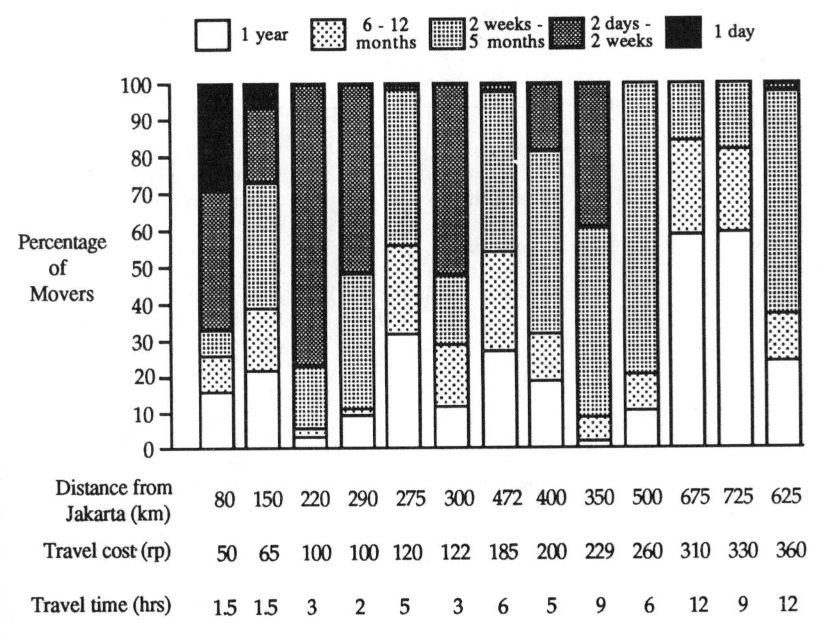

Figure 2.3 Circulation in West Java: the relationship between distance and duration of movement
Source: Graeme J. Hugo (1985) 'Circulation in West Java, Indonesia', in R. Mansell Prothero and Murray Chapman (eds), *Circulation in Third World Countries*, London: Routledge & Kegan Paul, p. 90.

within 65 miles, or 90 minutes travelling time, of Jakarta are there a number of migrants who commute daily to the capital city. Although the pattern is not entirely regular, the further away from Jakarta that villages are located, the greater is the volume of moves which span several weeks or months. It is important to note, however, that each village in the survey contained a variety of movement types in terms of the amount of time that people were absent from their home area.

As was the case with distance, Third World population movements also fall along a time continuum which ranges from, at one extreme, very short periods of absence (only a few hours) through to, at the other, movements which result in a permanent change in the migrant's usual place of residence. In between lie a wide range of time-spans over which different forms of movement spread. As with the spatial dimension, it is possible to boil these down to the few general types. These are outlined in Table 2.1.

Table 2.1 Temporal dimensions of population movement in the Third World

Time-span	*Type of movement*	*Characteristics*
SHORT-TERM		
A few hours	Oscillation	Farmwork; collecting (fuelwood, water)
Daily	Commuting	Journey to work, education, market
Weekly	Commuting	Away during the working week; entertainment, worship
Season	Seasonal circulation	Nomadism, pastoralism, transhumance; seasonal employment
Periodic	Sojourn	Hunting and gathering; trading, visiting
Once in a lifetime	Pilgrimage	Pilgrimage, marriage; displacement by natural disaster
Yearly	Contract labour migration	Target migration
Several years	Shifting cultivation	Shifting cultivation, frontier settlement
Working life	Temporary circulation	Urban-bound employment-related migration
Lifetime	Permanent migration	Emigration, resettlement, refugee movements
LONG-TERM		

At the short end of the time spectrum we find a variety of forms of movement which play a vital role in the daily lives of many people in the Third World. The land provides the main source of livelihood for some two-thirds of the Third World's population, and thus the daily movement of farmers to and from their fields represents one of the most prevalent forms of movement. In regions where the development of physical infrastructure is not particularly advanced, people may also spend several hours each day fetching water or collecting fuelwood. People who derive their livelihood from hunting, or gathering forest produce, also move regularly away from their home communities for short periods. Less frequent but no less important, particularly in areas where the capitalist economy has penetrated Third World societies, may be people's periodic trips to market to purchase provisions or to sell their produce. Mobility is thus of paramount importance to the peoples of the Third World. Where mobility is impaired by physical disability, poverty or inaccessibility, people may thus endure considerable hardship unless they can rely on the help of others within their communities.

As we saw earlier in this chapter, commuting is also an important form of short-term movement, which typically consists of people moving on a daily (or sometimes weekly) basis to work or study in another place, usually a nearby urban centre. Such a form of short-term movement is clearly only practicable where people enjoy relatively easy and inexpensive access to local destinations. Thus we find that the greatest level of work-related commuting in Third World countries tends to take place from communities which lie in close proximity to the larger towns and cities which are not only generally served by good transportation systems but which also offer the kinds of semi-skilled employment which may be appropriate for people who migrate from the countryside. People may also occasionally move to these centres on a less regular basis for reasons such as worship or entertainment.

In many parts of the Third World patterns of population movement may be determined by seasonal climatic patterns, especially in areas where there has been limited investment in irrigation facilities, for example, which might otherwise reduce the dependence of agriculture on seasonal variations in rainfall. As we shall see in the following chapter, the response may take the form of the periodic movement of people between ecological zones, as is the case with pastoralists and some semi-nomadic groups. Elsewhere, seasonal lulls in farming activity may prompt people to move in search of short-term employment with which to supplement their earnings from agriculture. Much, of course, depends upon the seasonal availability of employment opportunities both within and outside the local area. Such forms of movement are synchronized with the farming cycle, with people seldom staying away from their home communities for more than a few months (unless their prolonged absence is desirable from a financial viewpoint). A much longer cycle of movement is practised by shifting cultivators who, because of ecological restrictions on the length of time they can farm a given area, may only return to that area once every ten to twenty years.

Much employment-related migration in the Third World is not synchronized with the agricultural cycle, but instead involves periods of absence from the home community which can range from a few weeks through to the entire working lifetime of the migrant. It is important to draw a distinction between such temporary forms of circulation and those which are strictly seasonal because the effects of these different forms of circular movement, both on the place of origin and the place of destination, may vary quite significantly.

A key difficulty with identifying these longer-term forms of circulation is that the intentions of the mover with regard to the period of

Plate 2.1 Migrant samlor driver in Bangkok. Taxi-driving represents one of the more common forms of employment for migrants because it is relatively easy to get into, offers a reasonable livelihood, and allows seasonal migrants to come and go quite freely between the village and the city

absence from the home community may change quite radically during the course of his or her migration. Thus people may leave their home areas with the intention of returning within a year or two, only to find that their plans change on account of their fortune or ill-fortune in the chosen destination, because the migrant's personal or home circumstances have changed, or perhaps because the family they originally left behind decides to join the migrant in his or her new home. They may thus become unwilling, or occasionally unable, to return to their home communities and may settle permanently in the new location.

The duration of circulation at the longer-term end of the time continuum may thus only be identifiable *post facto* – that is after the migrant has returned to his or her natal area. Permanent migration can only, strictly speaking, be determined upon the death of the migrant, because whilst alive there is always a prospect of return, however remote. This is clearly unsatisfactory as a basis for differentiating permanent migration from long-term circulation and yet, from a planning viewpoint, it is often important to make a distinction between these two forms of population movement. As we shall see in Chapter 5, their respective effects on the places of origin and destination may be quite different.

Prothero and Chapman have argued that, whatever the actual duration of the movement, it is possible to differentiate between permanent migration and circulation by assessing the nature of the linkages, and psychological attachment, that the migrant maintains with his or her natal area. Thus where migrants retain property or land in the home community, remit money to family members, return periodically to visit relatives and friends, or maintain an interest and involvement in the political, social or cultural development of the home area, they might be considered to be preparing for their eventual return (whether or not they do in fact manage this). Such migrants may also invest much less of their time and money in setting down deep roots in the host community than may be the case with people who intend to settle permanently in the new location. Thus the *intentions* of the migrant at the time of departure with regard to whether or not he or she expects to return are an important basis for differentiating migration from circulation, notwithstanding the fact that migrants' intentions may change with the passage of time.

As we have seen, population movements in Third World countries fall into a wide range of categories depending upon the amount of time the mover spends (or intends to spend) away from his or her home community. These different types of movement may involve very different kinds of people in very different personal circumstances, or alternatively may involve people at different stages of their life cycle. Certain types of short-term movement may be particularly prevalent in less-developed regions where people are still very dependent upon nature and where there has been very little investment in infrastructure (irrigation, transportation, commercial marketing systems) which might help to overcome ecological and locational constraints on their livelihood. The wealthy may be better able to endure the extended absence of one or more household members, whereas the poor may be obliged to enter into a pattern of seasonal circulation which allows them to maximize their returns from farming and employment elsewhere. The migration of entire family groups, or unmarried youths, may be more likely to result in their permanent relocation than will be the case with individual migrants whose spouses and families remain in the place of origin.

The temporal dimension of population movements in the Third

World is therefore vitally important, partly because it tells us a great deal about the circumstances which may underpin the decision to migrate, and partly because it has a significant bearing upon the scale of movement which may be identified. The tendency for several Third World governments to define population movements for official planning purposes as involving a long-term or permanent separation of people from their home communities means that a significant proportion of the types of movement discussed above are overlooked. Not only are these governments under-recording the total volume of movement within their respective countries, they are also overlooking a significant variety of movements which reveal a great deal of information about the characteristics and problems of development within their national territories.

The motivational dimension

The factors which motivate people to leave their home communities provide another basis for classifying types of population movement in the Third World. The most important distinction to be made is that between voluntary and involuntary forms of movement – that is, between cases where the mover has a more or less free choice concerning whether or not to leave and those where the mover has no such freedom of choice. We have already seen, however, that it is often rather difficult to make a clear distinction between *voluntary* and *involuntary* movements. Certain forms of movement, such as rural–urban migration, may appear to be the result of free choice, but the circumstances which faced the migrant may in reality have left little option but to move. The choice may have been between survival and starvation. Conversely, the decision of refugees to flee their home areas, the archetypal form of involuntary movement, may in part be rationalized by the expectation not only of asylum but also of much better economic prospects in a new country.

Figure 2.4 overcomes this difficulty by including a third category of *impelled* movements, where the mover retains a modicum of power to decide whether or not to leave. Political refugees may be said to fall into this category, in that they have a certain amount of choice about leaving or staying and about when to depart should they so choose. The fact that people have different levels of tolerance of, or are affected to different degrees by, political persecution determines that some people decide sooner than others that it would be in their best interests to

	Conservative	*Innovative*	
Forced	Displacement (evacuees)	Slave trade	
Impelled	Flight (refugees); traditional (rites of passage); ecological (shifting cultivation, pastoralism)	Coolie trade, indentured labour	*Dependent*
Free	Migration ('push' factors predominant)	Migration ('pull' factors predominant)	*Independent*

Figure 2.4 Motivational dimensions of population movement in the Third World

Source: Adapted from W. Petersen (1958) 'A general typology of migration', *American Sociological Review* 23 (3), p. 266.

leave. Similarly, different people give different weight to the advantages of moving to a new place set against the disadvantages of leaving their home areas.

Some observers draw a distinction between 'anticipatory' and 'acute' refugee movements. Although in both cases the refugee would ideally prefer not to have to move, in the former instance refugees are able to prepare in advance to leave their country of origin, taking their possessions with them and exercizing a modicum of choice over where to move to, whereas for the latter the decision to leave is often a spontaneous one, leaving little or no time to make proper arrangements. Whilst the former may, superficially, be difficult to differentiate from voluntary forms of migration, the latter is clearly a form of involuntary movement, with the emphasis being placed on the escape from danger with little immediate concern being given to the longer-term consequences of such a course of action.

Several other forms of movement which will be described in the following chapter (e.g. ecologically or culturally determined movements) might also be categorized as 'impelled' in that the people who move do at least have a theoretical option of remaining where they are.

A distinction is also made in Figure 2.4 between *independent* and *dependent* movements. In the former case the decision to move to a new location is the prerogative of the individual, whereas the latter refers to movements where the decision has either been taken collectively (e.g. by the mover's family) or has been taken by someone other than the mover. We will see in later chapters that, in many parts of the Third World, the act of migration is quite commonly used as a means of extending or diversifying the household economy beyond the confines of the home area. In such cases the decision to migrate may not be made by the mover alone, but may involve other (usually more senior) household or community members. The move is then made in support of the decision-making unit as a whole rather than for the exclusive benefit of the mover. The implications of these different forms of migration decision-making will be discussed in Chapter 5.

Finally, a distinction is made between *conservative* and *innovative* forms of movement. In the former case the main motivation for movement is for people to preserve the style or standard of living that they already have, whereas in the latter people migrate in order to achieve something new. In Figure 2.4 it is suggested that situations where people leave their home communities in response to poor or deteriorating conditions therein ('push' factors – see Chapter 4) would constitute conservative forms of migration, whereas cases where people leave in response to the lure of opportunities and better prospects elsewhere ('pull' factors) represent innovative forms of movement. Who the prime decision-maker is, and the associated reasons for the movement may have a significant bearing on both the characteristics (e.g. timing and duration) and the effects (e.g. money remitted) of the move.

As we shall see in later chapters, the overriding motivation for many forms of population movement is provided by economic factors. People leave their home areas because local economic opportunities are inadequate to satisfy their needs and aspirations and because opportunities elsewhere are perceived to offer better longer-term prospects. Linked in with the economic motivation may be various forms of social motivation. Thus the mover may expect to enhance his or her standin in the local community as a result of the financial returns and/or the social prestige which may be bestowed upon people who spend a period living and working away from the home community. Conversely, people may be forced into migration by social conditions in their home communities, such as conflict with other families or some form of scandal or loss of face.

Conclusion

The aim of this chapter has been to attempt a simple categorization of the many and varied forms of population movement which occur in Third World countries. Whilst it has been possible to identify some of the more important common denominators – such as duration, direction, distance and decision-making – it should be clear that even in their more simplified or streamlined form the characteristics of movement none the less remain very complex. Whilst it may be possible to pick out certain trends and patterns, there is no single form of movement which may be considered typical of particular people in particular circumstances at particular moments in time. People are notoriously fickle, and seldom consistent, in the way that they respond to prevailing circumstances – and this is a trait which is not unique to the Third World. No two people can be expected to respond in an identical way to what may appear to be identical circumstances and milieux. Similarly, the way that people respond, and the circumstances to which they respond, are constantly changing, making the categorization and prediction of movement types an extremely difficult and dangerous task.

This chapter has identified some of the more important variables which are involved in the processes of decision-making, movement and return. These will be analysed in more detail in Chapter 4. However, we will first switch from the general to the specific, using our simple typology to identify and examine a variety of forms of population movement which are found in the Third World today.

Key ideas

1 The basic elements of population mobility – space, time and purpose – are much the same the world over. Where the Third World differs from other parts of the world is in the developmental, social and cultural environment within which movements occur.

2 The spatial pattern of many forms of population movement in Third World countries mirrors closely the characteristics of the development process.
3 Circulation is often the most prevalent type of movement in Third World countries. Short-term, circular and cyclical forms of migration represent a means by which people can take advantage of opportunities available elsewhere whilst retaining their stake and position in their home communities.
4 Economic forces exert the strongest influence on the mobility decisions of people in the Third World but social, cultural, political and environmental factors may determine the characteristics and impact of the resultant movement.
5 Each year several million people initiate or conclude a movement from one place to another, and each movement is in its own way unique. The search for patterned regularities in the process of movement is essential for analysis and policy-making, but serves only to simplify and generalize what is a hugely complex and complicated process.

3
Forms of population movement in the Third World

In what ways, if any, do population movements in the Third World differ from those which take place in the industrialized and economically advanced countries of the world? Although we should be mindful about the dangers of viewing the Third World as a separate and homogeneous entity, it is possible to identify some basic differences in the spatial, temporal and motivational characteristics of movement which occur in different parts of the world. In general, these differences can be explained by the contrasting economic, social, cultural, historical and environmental contexts within which they take place.

The aim of this chapter is to introduce a variety of forms of population movement which, to a greater or lesser extent, also highlight some of the distinguishing features of Third World regions. None however is the 'exclusive' domain of the Third World. Several of the forms of movement which are described below have counterparts in the Developed World. Thus for example the periodic movement of pastoralists between different ecological zones in the Himalayas and Andes is similar in many respects to the transhumance which is still widely practised in Alpine and other upland regions of Western Europe. The pilgrimage of Muslims to Mecca or of Hindus to the Ganges is also paralleled by that of Christians to Lourdes.

Traditional forms of population movement

Throughout history, people have moved in search of land, food, employment, adventure, and even spouses. In some areas mobility plays such an important role in people's lives that the entire structure of society has been built around the periodic movement of groups of people. Such is the case with nomads, hunter-gatherers, shifting cultivators and pastoralists for whom movement forms an integral part of their way of life. Such traditional forms of population movement continue to the present day, little different in character and function to those dating back perhaps several centuries. Their resilience to the forces of modernization and development is explained by the fact that they tend to occur in areas which are somewhat isolated from the main centres of economic activity and political decision-making. In other societies, the difficulties and dangers associated with venturing into the unfamiliar and challenging world outside the home community have led to migration being seen as a stern test of adulthood from which few young males would wish to shy away. Such traditions of movement may in turn have enabled people to adapt more readily to change in their wider societies.

It is tempting to view such traditions of movement as a 'residual' form of activity – a remnant of a bygone era which is rapidly disappearing in the face of more powerful and pervasive forces of development and modernization. Certainly, the numbers of people involved may rapidly be dwindling, but they remain important, not least because they provide a valuable insight into the resilience and adaptability of traditional Third World societies. The following discussion looks at the role of culture and ecology in underpinning traditional forms of movement.

Culturally-determined movements

From adolescent to adult

Population movements are interwoven with the cultural fabric of some African and Asian tribal societies. Historically, the movement of young males from their home communities, often for several months at a time and covering large distances, represented a traditional means by which they demonstrated their bravery and prowess and their preparedness for the transition from adolescence to manhood. Such movements may typically have occurred in association with tribal warfare or hunting expeditions, or may have taken the form of individuals demonstrating their ability to survive, alone, in hostile territory.

Whilst labour migration may have replaced hunting or going to battle as the predominant reason for out-migration today, a great deal of cultural and social significance may none the less still be attached to the periodic movement of people to find work. A study of the Gcaleka tribe in the Transkei region of South Africa has shown how the ritual events which accompany a migrant's departure from and return to his home community closely parallel those which traditionally accompanied tribal initiation rituals. Among the Tswana and Basutho tribes of southern Africa, there is also evidence that labour migration is regarded as an integral part of an adolescent male's initiation into manhood. The boy becomes a man in the context of labour migration: he becomes a provider and ceases being a dependant. Migration is a test: those who pass it demonstrate the maturity to be able to survive the difficulties and dangers of urban life, and to accept responsibility.

Whether or not migration should be seen as part of the initiation into adulthood is strongly debated, as is the notion that these traditions of movement may have laid the foundation for contemporary forms of migration. What is clear, however, is that the movement between different economic, social and cultural spheres which is associated with migration often has a powerful impact upon the people involved, and that in many societies a certain amount of prestige and status accrues to those who return successfully from their sojourn elsewhere. In the north-east of Thailand a tradition of *pay thiaw* (literally 'to go wandering', always with an overtone of fun and enjoyment) involving the movement of small groups of young men, often with no specific destination or timetable, can be traced back over several centuries. The strong cultural value which is attached to mobility, and to the knowledge and experience of different areas, meant that travellers generally could count on enhanced social status upon their return to their home communities. Conversely, young men were regarded as ignorant and cowardly if they had never moved away from the village. *Pay thiaw*, as with other traditional forms of mobility, represented a form of circular migration: failure to return was considered tantamount to deserting one's parents and evading one's responsibilities.

The Minangkabau in Indonesia have a tradition of *merantau* whereby young men leave their home communities in order to obtain knowledge and experience before returning home. Even today men are still encouraged to migrate before marriage in order to prove to themselves and others that they are capable of earning a livelihood and standing on their own two feet. The Iban of Sarawak in East Malaysia have a traditional custom of *bejalai* (literally 'going around'), where young unmarried men leave their longhouses to undertake journeys which may be expected to bring them material gain or social prestige. The roots of the *bejalai* tradition lie in the legendary tales of Iban culture heroes who experienced wild adventures during their years of wandering. One such hero, Keling, remains the ideal of Iban manhood, and this explains not only why the tradition of 'going around' continues to the present day, but also why Ibans who return from their journeys are proud to relate their own adventures (customarily laced with imaginative embellishments).

Bejalai was often closely associated with warfare and raiding, including headhunting, and with trips to collect forest produce and to engage in trade. Traditionally, *bejalai* was undertaken by groups of men rather than individuals, and it was expected that young Iban males would undertake *bejalai* at least once in their lives. Considerable social pressure was placed on young Ibans to follow in the footsteps of their forebears. During the present century the Iban have adapted the tradition of *bejalai* to include labour migration to urban centres in Sarawak and further afield.

Pilgrimage

The central role that religion plays in the lives of many Third World societies is reflected in the large volumes of people who occasionally travel considerable distances on pilgrimages to sacred places. Muslims are obliged to undertake the *haj*, or pilgrimage to Mecca, at least once in their lives, provided they are physically and financially able to do so. Devout adherents of other religions may make pilgrimages for less formally prescribed reasons, such as the fulfilment of a vow, the reaffirmation of one's faith, or the search of a cure for ill health.

In spite of the forces of modernization and Westernization, which in some societies have tended to undermine the strength and role of religion, particularly among the younger generations, we find that the numbers of people involved in pilgrimages have risen steadily in recent decades. This paradox is explained, on the one hand, by widespread improvements in transportation and communications systems, and in the facilities available at the major sacred sites, and on the other by the growth of fanaticism and fervour amongst certain religious elements and sects.

For most, pilgrimage may involve only a relatively short-distance, short-term movement to a local shrine or religious centre with significance to the beliefs and practices of particular groups. The volume of movement may be relatively small, and spread more or less evenly throughout the year. Much larger volumes of movement may occur in conjunction with major religious festivals at certain times of the year or, with the Hindu *kumbha* bathing festival which attracts between 1 and 5 million people to the banks of the river Ganges, once every twelve years. As many as 21 million individual pilgrimages may be undertaken in India each year, mostly to the shrines of deities which are of local or

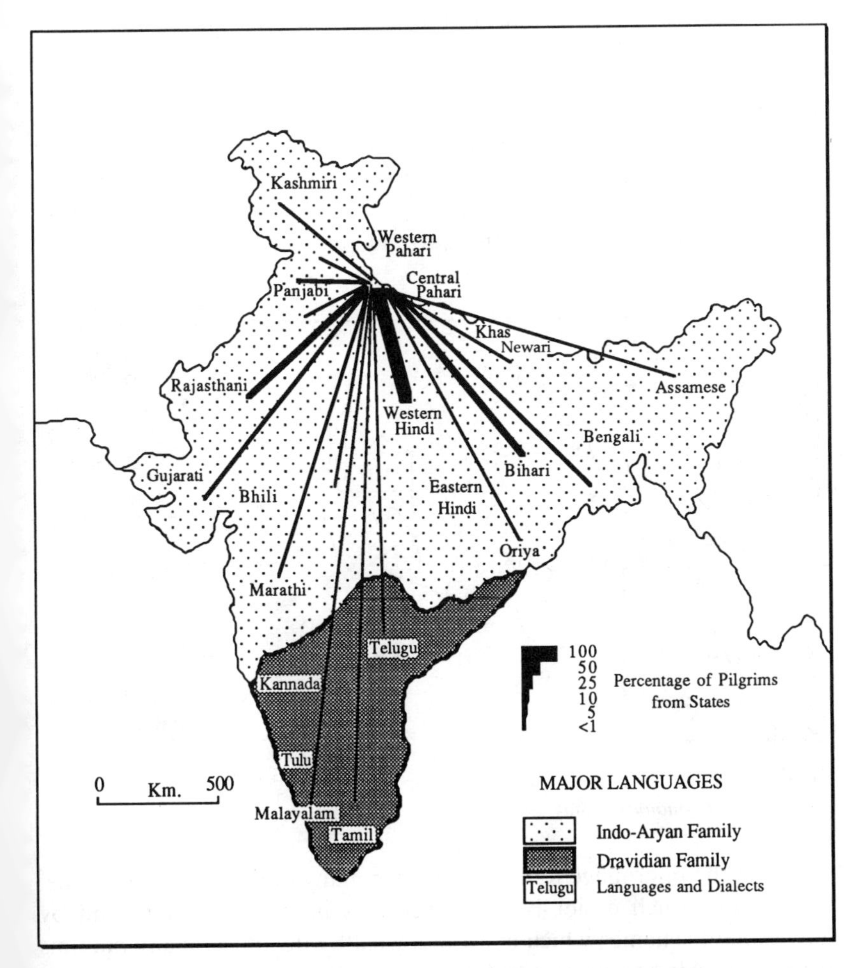

Figure 3.1 Areas of origin of pilgrims to the Badrinavain 1968 Yatra
Source: R. Mansell Prothero and Murray Chapman (eds) (1985) *Circulation in Third World Countries*, London: Routledge & Kegan Paul, p. 260.

regional importance. Such pilgrimages may be undertaken several times during a person's lifetime and, because of ethnic and linguistic variations, may predominantly occur over relatively short distances, as Figure 3.1 clearly illustrates.

The *haj*, by contrast, mostly involves international and intercontinental movements which the pilgrim is usually only able to undertake once in his or her lifetime. Up to 2 million people converge on Mecca and Medina in Saudi Arabia during the month of pilgrimage (see Figure 3.2), at least half of whom come from more than 130 countries worldwide stretching between Indonesia, which has the world's largest Muslim population, West Africa and beyond. Mecca holds a central place in the lives of all Muslims. It was in Mecca that the Prophet Mohammed was born, and it was here that he first heard God's call. Pilgrimage to Mecca is one of the five central duties of devout Muslims, and has been likened to returning to the very foundation of the faith.

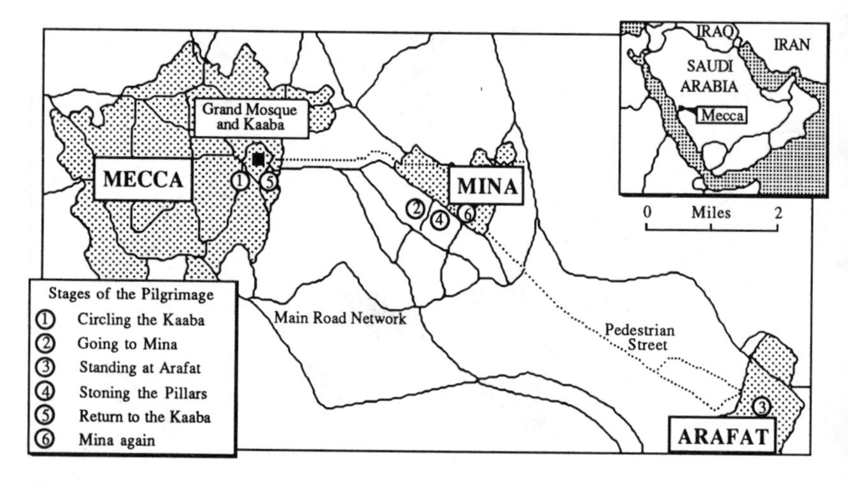

Figure 3.2 Mecca: place of pilgrimage for the world's Muslims
Source: The Economist, 7 July 1990, p. 65.

Whilst the pilgrimage to Mecca has, for many of the world's Muslims, been made much easier by improvements in transportation, and by government schemes which assist less wealthy devotees, many pilgrims from West Africa continue to follow traditional routes of pilgrimage to Mecca which may take several years to complete. Some 3–4,000 pilgrims each year embark from Nigeria and Niger in the general

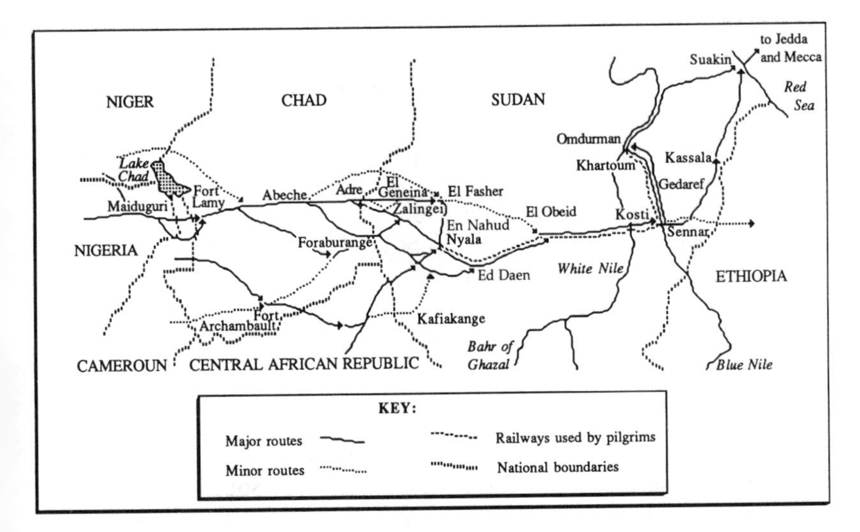

Figure 3.3 Routes of overland pilgrimage by West Africans to Mecca
Source: J. Stace Birks (1975) 'Overland pilgrimage in the savanna lands of Africa', in L. A. Kosinski and R. M. Prothero (eds), *People on the Move: Studies on Internal Migration*, London: Methuen, pp. 297–307.

direction of Mecca, following a route through the northern savannas (see Figure 3.3) which their ancestors have travelled since at least the eleventh century. Scattered along the route are little communities of West Africans (mostly Hausa) who provide shelter and work for the small groups of pilgrims, and information concerning the route to be followed. The pilgrims will, in effect, 'work their passage', and have also to save sufficient funds to enable them to make the final stage of the journey to Mecca. Because of the long and difficult overland journey, and the pilgrims' need to work to support themselves, the pilgrimage to Mecca may take anything between three and twenty-five years to complete. Once in Mecca they may then rub shoulders with Muslims who have enjoyed first class air travel to Saudi Arabia and who stay in luxury air-conditioned hotels.

Ecologically determined movements

It is a feature of many parts of the Third World, especially in the more remote and ecologically marginal areas, that human activity is to a large extent controlled by nature. Dispersed and isolated societies often lack technological skills which would enable them to exert an element of control over the natural environment. Furthermore, the ecological conditions with which many in such areas are faced, particularly in arid, mountainous, coastal and jungle regions, may provide a formidable challenge to anyone wishing to exploit the natural environment for agricultural or other economic purposes. People in these marginal and remote areas have thus come to rely for their livelihood upon what nature grudgingly provides. The response of many Third World societies to environmental constraints on their livelihood has been to move periodically between different areas and ecological zones. Some of the more prevalent forms of movement will be discussed briefly below.

Hunting and gathering

Hunter-gatherers, such as the Bushmen of the Kalahari and the Ona and Yahgan Indians in the southernmost parts of South America, may cover vast areas of land as they track wild animals or search for wild fruits, berries and other natural produce. Although movement is very much a part of their way of life, the distance and direction of their movement is closely associated with prevailing ecological, and especially climatic, conditions. The Agta Negritos, in north-eastern Philippines, move camp every seven to ten days during the long dry season when food becomes too scarce to support their small communities for an extended period in one place. In recent years the Agta have also had to contend with the steady encroachment of technologically more advanced sedentary farmers into the territory they have traditionally occupied. As a result of growing competition for the natural resources which have sustained them for centuries, the Agta are increasingly unable to maintain their free-ranging existence. At the same time, their growing need for cash and the increasing opportunities that now exist to engage in paid employment for local sedentary farmers have combined to encourage a growing number of Agta to adopt a more settled lifestyle. For the remainder, the semi-nomadic way of life remains deeply ingrained in their traditional culture.

Shifting cultivation

Shifting cultivation (swiddening) represents another situation wherein people regularly move from site to site, clearing small tracts of land every 1–3 years, because prevailing ecological conditions will not support permanent forms of agriculture (at least with prevailing levels

Plate 3.1 Tree burning and land clearance by Iban shifting cultivators in Sarawak, East Malaysia

of technology). Because of thin, infertile soils yields decline sharply after a couple of years of cultivation, forcing swiddeners to move to another location, leaving their land to lie fallow for up to twenty years before it can be cultivated again.

Swidden farmers, such as those found in the upland regions of South-East Asia, demonstrate a high level of dependence on nature, but also characteristically live in close harmony with the natural environment, demonstrating a very sound understanding of ecological principles. Movement from one site to another is an integral part of their balanced economic system. Because of the fragility of the ecosystems within which they operate, they cannot afford to over-exploit the natural resources upon which they depend. In many parts of South-East Asia, the expansion of commercial logging and the growth of population have combined to exert excessive pressure on these resources, with the effect that cultivation cycles, and with them cycles of population movement, have become greatly modified. In some areas, farmers have been forced to turn to sedentary forms of agriculture.

Nomadism

Although its overall presence is small and declining, nomadism can be identified in many parts of the contemporary Third World. Its characteristics vary. In the southern tip of South America, a few surviving Chono, Alacaluf and Yaghan Indians engage in nomadic shellfish gathering, as their forebears have done for centuries. Travelling by canoe, they ply the beaches and islands of the Chilean archipelago gathering shellfish, and searching for the occasional stranded seal or whale. They have traditionally turned for their livelihood to a roaming existence at sea because inland the terrain is difficult, and does not provide sufficient game and plants to satisfy their dietary needs.

In the Malayan archipelago (Malaysia, Singapore and Indonesia), the so-called 'sea gypsies' (*orang laut*) provide another interesting example of a boat-dwelling nomadic people. The *orang laut* are restricted to the more isolated and marginal areas of the archipelago, such as rugged small islands, mangrove-clad shorelines and swampy coastal lowlands. Population densities are very low, due mainly to the fact that their principal source of livelihood is derived from the collection of marine and forest produce, which may involve covering vast tracts of land and ocean. The kind of environment within which they operate and the level of technology they employ prevents the development of large concentrations of population. Typically, the *orang laut* will move around in small groups, camping and fishing together, with their boats serving as the family home. As with the South American Indians, many *orang laut* have turned to sedentary sources of livelihood because of growing competition for the territory and products upon which they rely, and as a result of their closer association with permanent agriculturalists and the modern world in general.

On dry land the Bedouin, whose traditional homeland is the Arabian Desert (bordering Iraq, Syria and Jordan), are one of the most notorious of all nomadic groups. The extent of their nomadism varies according to the time of year: they often spend the summer months camped around wells or streams, and the rest of the year ranging the desert. The same is true of other nomadic groups (see Figure 3.4). Their chief possessions are their camels and their home – the diagnostic Bedouin tent, a long, low black tent woven out of goat hair. Raiding provided a traditional means by which the Bedouin supplemented the deficiencies of life in an arid zone, which has added to their fearsome reputation. Indeed, raiding and piracy have historically been quite common amongst nomadic groups in general, reflecting not only their

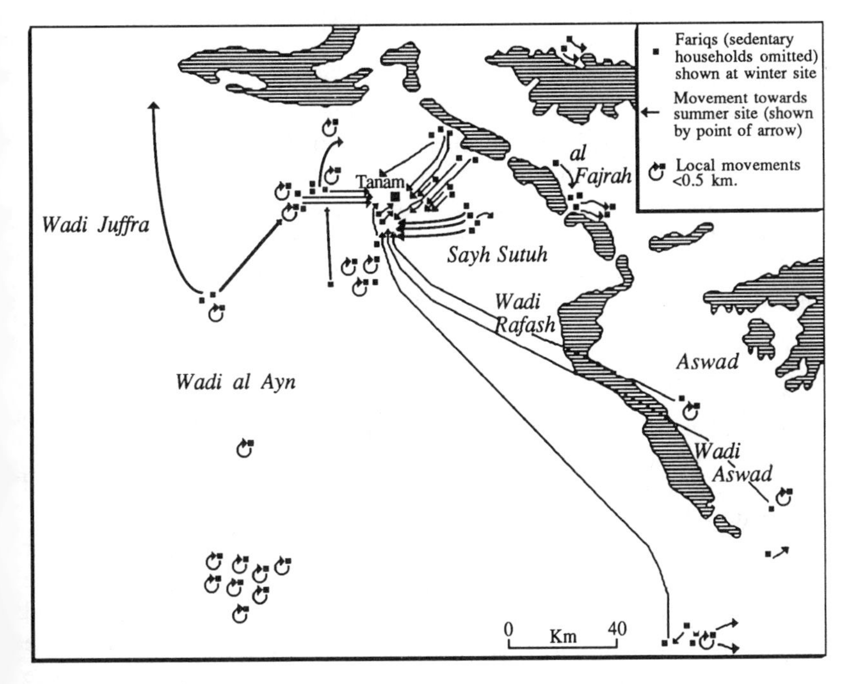

Figure 3.4 Seasonal pattern of movement by nomadic *Fariqs* in Saudi Arabia
Source: J. Stace Birks (1985) 'Traditional and modern patterns of circulation of pastoral nomads: the Duru of south–east Arabia', in R. Mansell Prothero and Murray Chapman (eds), *Circulation in Third World Countries*, London: Routledge & Kegan Paul, p. 157.

frequent inability to satisfy all their consumption needs by collecting the produce of nature, but also the tough and resilient breed of people which the nomadic way of life appears to engender. In recent decades many Bedouin have swapped their nomadic lifestyle in favour of migration to urban areas to work in the oil industry, or for permanent settlement on government-sponsored tracts of land.

Elsewhere in the Third World the phenomenon of nomadism is also declining steadily, for a number of reasons. First, nomadic groups are being displaced, or absorbed, by more advanced agricultural (and industrial) societies. Second, national governments have had a significant influence on the decline of nomadism, ever fearful of the security implications of allowing groups of people to roam freely across national boundaries. Programmes of sedentarization or resettlement, as are firmly promoted in Syria, are commonly justified on the grounds of improving the social welfare of nomadic groups. Third, nomadic societies are coming under increasing pressure as a result of competition for land (particularly in areas where land has some potential for alternative uses). Their lack of legal rights over the territory that they have traditionally occupied for centuries has heightened their vulnerability in this regard. It is not only human pressures which are affecting nomadic groups, however. Tuareg nomads from southern Algeria, Mauretania, Niger and Mali have been displaced by the shifting sands of the Sahara, and now several thousands of them eke out a meagre living by begging in the capital cities of their respective countries.

Transhumance

A final illustration of population movements which are closely associated with ecological factors concerns transhumance or semi-nomadic pastoralism, which is prevalent in the more mountainous parts of the Third World. Transhumance involves the periodic shift of small numbers of population between different environmental zones (typically upland and lowland) in response to prevailing climatic and ecological conditions. Transhumance is widely practised in the mountainous regions of the South Asian sub-continent. Approximately one-fifth of the land area of Tibet is roamed by nomadic pastoralists. In the Swat region of Pakistan, semi-nomadic Kohistanis move annually between altitude belts, ranging from 2,000–14,000 feet. Because of prevailing ecological conditions in north-western India it is not possible to graze livestock year-round in one area. With the onset of the monsoon, livestock are moved from Gujarat towards the sub-mountainous regions of Rajasthan. The vacated lowland fields are then used to grow arable crops. After the harvest, around December, the herds return to graze in the lowlands.

In many parts of South Asia the practice of transhumance is steadily declining. Pressures are mounting as a result of growing competition for land in upland areas, and growing shortages of labour for livestock herding and lowland farming. Because they often lack formal, legally defined access to land semi-nomadic groups also tend to be in a weak position when faced with competition for land from other sources. The result is a tendency towards permanent settlement where conditions or production systems allow. We also tend to find a decline in transhumance where improvements in transportation and communications have reduced people's reliance on the goods provided by these semi-nomadic groups.

Involuntary population movements

The traditional forms of movement described above would, in the main, fall into the 'impelled movement' category which was outlined in Chapter 2. Whilst in most cases there is little realistic alternative to movement, so powerful are the respective influences of environment and tradition, periodic shifts in the place of residence of several Third World societies are so deeply ingrained in the lives and cultures of those involved that it would be misleading to view them in the same context as the 'involuntary' forms of movement described below.

The latter take a wide variety of forms and occur in response to a wide range of circumstances, including escape from warfare, civil conflict, revolution, discrimination, religious rivalry, natural disasters and the displacement of people by development programmes. Such difficulties have, over the millennia, forced people from their homes throughout the world. However, in recent decades involuntary movements have become particularly prevalent in the Third World. Several factors help to explain this situation. First, during the post-Second World War period, the Third World has played host (often with superpower influence) to some of the world's most violent and protracted conflicts which have been driven by clashes of ideology, ethnic interests, political factionalism and dictatorial regimes. As a result, millions of people in Central America, South-East Asia, the Middle East and much of the continent of Africa have been displaced from their homes by fighting, associated economic disruption, abuses of human rights and/or political persecution.

Second, many parts of the Third World lie in regions of the globe which are particularly prone to natural disasters such as floods, drought, earthquakes, volcanic eruptions, pest infestations and so on, which, on occasions, have forced many thousands of people from their homes, sometimes permanently and tragically. Whether or not global environmental changes are responsible for the increased frequency and intensity of such natural events, it appears that the countries of the Third World are less able to afford the infrastructural and technological investments which are necessary to prevent or ameliorate the human consequences of natural disasters. Third, the pace and pattern of economic development in many Third World countries has been such that increasingly large numbers of people have been displaced by major infrastructural projects and by the commercial sector's voracious appetite for land. Economic growth is not only responsible for the growing number and scale of such changes, but the relative financial and political weakness of scattered communities leaves them powerless to resist the relentless forces of capitalism and development.

The following discussion will focus briefly on two forms of involuntary population movement in the Third World.

Refugees

The scale of movement of people seeking refuge from political (and ecological) crises has expanded rapidly over time as populations have grown: it is calculated that some 140 million people have been displaced during this century alone. Furthermore, since the 1950s some 95 per cent of refugee movements have originated within the Third World. Table 3.1 shows that there were an estimated 14 million refugees in 1988 alone, including almost 6 million Afghan refugees in Pakistan and Iran. The true figure may be much greater. A more realistic total for Africa, for example, would be in excess of 6 million refugees. To this must be added some 9 million 'internally displaced persons' (people who have been displaced from their homes, but who remain within their country of origin) in Africa (see Figure 3.5), and more than 1 million in El Salvador and Guatemala. More recently, perhaps as many as 2 million foreign workers, mostly Asian and North African, became temporary refugees with the onset of the Gulf Crisis in late 1990. Since the conclusion of the conflict, almost 1 million Yemenites have been expelled from Saudi Arabia.

Table 3.1 Refugees in need of protection and/or assistance, 1988

Asylum region	*Total*
Africa	4,088,260
East Asia and the Pacific	625,780
Europe	356,000
Latin America and the Caribbean	279,850
Middle East and South Asia	9,071,910
Total	14,421,800

Source: United States Committee for Refugees (1989) *World Refugee Survey, 1988*, Washington, DC, pp. 32–3.

The continent of Africa features some of the most hopeless and depressing aspects of the world's refugee crisis. The scale of population displacement increased fivefold between 1981–6, and the United Nations claims that up to 100 million people in Africa are at risk from the political and environmental factors which have already created large

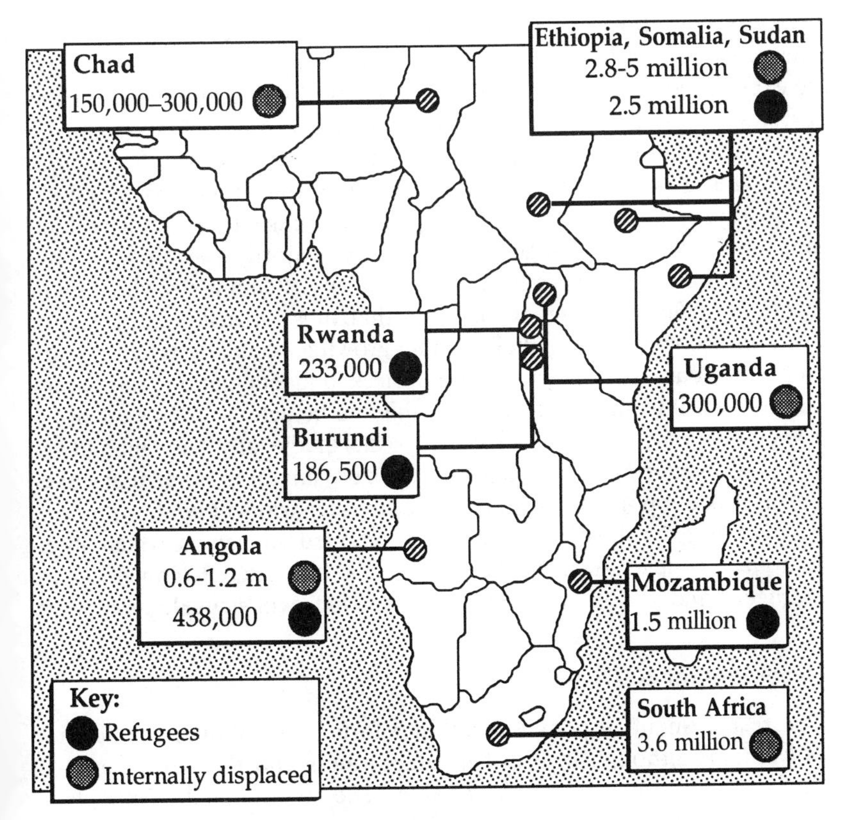

Figure 3.5 Examples of numbers of refugees and 'internally displaced persons' in Africa
Source: The Guardian (World Media), 14 June 1991, p. 27.

numbers of refugees. Drought and desertification in the Sahel continue to displace hundreds of thousands of people, many of them severely malnourished. Military conflict, ethnic rivalry and economic under-development determine that they meet great difficulty and hardship in finding a suitable place of refuge. The countries which have traditionally provided refuge are themselves facing severe economic and resource problems. Some three-quarters of African refugees are women and children (reflecting general demographic patterns) who are particularly vulnerable to the many problems which are associated with their flight from the comfort and familiarity of their home communities.

Although refugee movements were categorized in Chapter 2 as

'conservative' forms of movement, where the mover seeks to protect that which she or he already possesses, in reality the refugee can expect to experience a significant deterioration in well-being, both material and psychological, particularly in the short-term, even if successful in gaining asylum in a second country. None the less, most refugees hold on to the prospect of eventually returning to their home communities. In reality, it seems probable that, given the present nature of political and environmental change in Africa, the majority of refugee movements will in future necessitate permanent resettlement elsewhere in the continent. Given the great difficulty that most African states already have in accommodating their rapidly-growing populations, with rising debt burdens, falling commodity prices and stagnating agricultural economies, it is difficult to see how this can be achieved. In view of the reluctance and, given the projected scale of the problem, inability of the world's more affluent countries to make anything other than token gestures of support, the most likely future scenario will be the sprouting of ever-larger refugee populations contained in camps which international agencies such as the United Nations High Commission for Refugees and various charitable organizations will struggle to support.

Resettlement

The resettlement of people is another form of involuntary population movement, although levels of involuntariness and the permanence of the movement may be quite variable. The resettlement of people to make way for major infrastructural projects (e.g. ports, transportation termini, or reservoirs for irrigation and power generation) involves the enforced and permanent movement of people from one site to another, to which there is no practical alternative. The redistribution of population may involve compulsory measures, such as the eviction of squatters in Lima or Manila in the 1960s and 1970s and the consolidation of ethnic minority groups in Vietnam, or may include an element of volition on the part of the mover, as with various land settlement schemes in Indonesia and Malaysia. In most cases, however, resettlement is a form of involuntary population movement because, given the choice, the movers would generally have preferred to stay put.

Over the last three to four decades the Third World has played host to some quite massive reservoir construction projects, many of which have been funded by international agencies such as the World Bank. Table 3.2 illustrates the huge scale of several of these projects. Although

Case study A

The Vietnamese 'boat people': economic migrants or political refugees?

On 29 October 1991 Vietnam, Hong Kong and Great Britain signed an accord which opened the way for the forced repatriation of so-called 'boat people' to Vietnam. Up to 50,000 people now face an involuntary return to the country that they fled, in many cases with nothing but the clothes on their backs, having got no further than the squalid conditions of a Hong Kong detention centre. The enforced return of fifty-one Vietnamese asylum seekers in December 1989 caused an international outcry and was followed by riots in the detention camps. This time, the detainees have pledged mass suicides, proclaiming 'Dead is better than Red'.

At the heart of their plight lies the narrow and controversial distinction between a 'political refugee' and an 'economic migrant'. Until mid-1979, there was little challenge to the refugee status of the more than half a million Vietnamese and ethnic Chinese who fled the Communist takeover in Vietnam. Many had worked closely with the Americans before the South fell to the Communists in April 1975, and were readily accepted for resettlement in the United States. Subsequent waves of refugees left because of political persecution, the Communist regime's firm clampdown on the capitalist sector, and because of Vietnam's deteriorating relationship with China, which particularly affected the safety and economic well-being of the ethnic Chinese community in Vietnam. A large proportion of departees were Vietnamese Chinese.

The vast majority of people who have left Vietnam since 1979 have been ethnic Vietnamese, drawn mainly from the lower and middle classes – people who do not appear to have been the direct targets of state repression. Their departure has occurred at a time of economic crisis in Vietnam, and has led to the assumption that these people are leaving Vietnam because of the prospect of a better life after resettlement in a new country. There is little doubt that economic factors have figured quite prominently in the movement decisions of many, but they do not tell the whole story. Many have also fled Vietnam because they face discrimination, fear persecution, or because they are ideologically opposed to the

Case study A *(continued)*

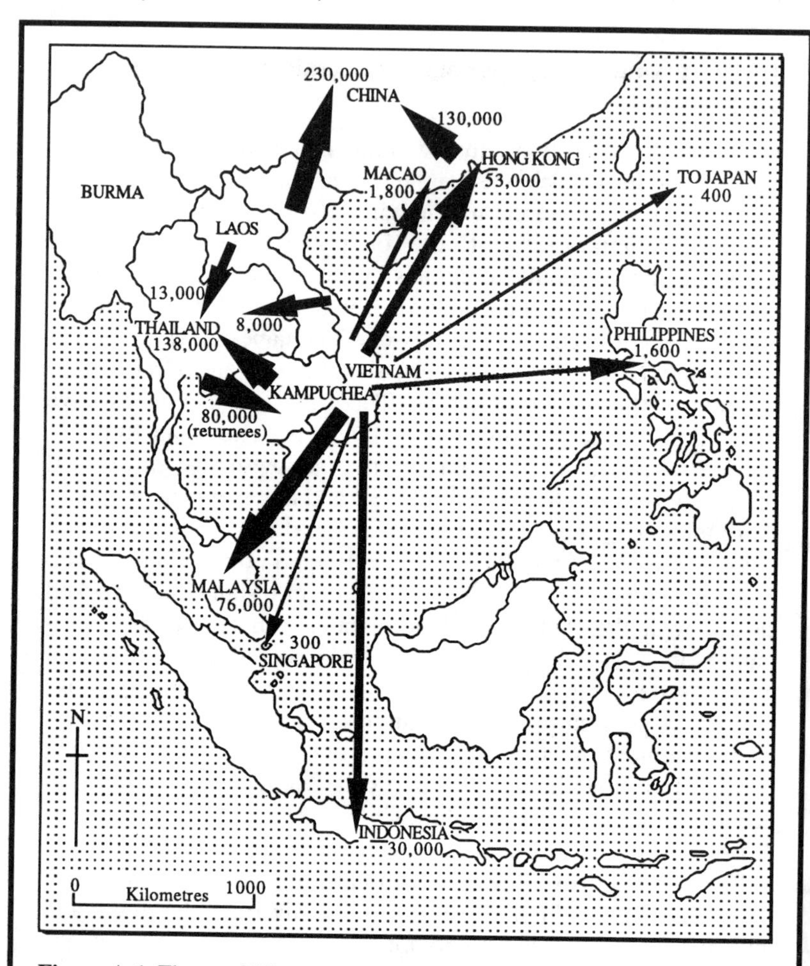

Figure A.1 Flows of Vietnamese refugees in South-East Asia, 1975–9 (excluding those resettled elsewhere)
Source: G. J. Lewis (1982) *Human Migration: A Geographical Perspective,* London: Croom Helm, p. 38.

Communist regime. In most cases the economic and political motives for flight are inextricably linked.

The authorities in Hong Kong and other countries of first asylum (e.g. Malaysia, the Philippines, Indonesia, Thailand) have

Case study A *(continued)*

emphasized the economic basis of the post-1979 movements and have sought to screen out the economic opportunists from the genuine political refugees. The narrowing interpretation of a 'refugee' has coincided with the 'compassion fatigue' which has afflicted several Western nations and which has led to drastically reduced resettlement quotas for Vietnamese refugees. The large numbers of Vietnamese (and also Laotian and Cambodian) refugees in camps throughout South-East Asia is further testament to the international community's reluctance, or inability, to deal with the refugee problem by accepting people for resettlement.

Whatever the reasons for the screening out of the economic migrants, it will shortly be used as a powerful tool for clearing Hong Kong's detention centres of a large number of unwanted and unfortunate boat people. In spite of assurances from Vietnam that returnees will be well treated, the government may none the less be expected to encounter considerable difficulty in absorbing such large numbers of people given the precarious state of the Vietnamese economy. The recent resolution of the 'Cambodia problem', and the continuing *rapprochement* between Vietnam and the United States might augur well for the longer term, but the immediate prospects for those who return under the forced repatriation programme would appear to be quite bleak.

Source: Andrew Johnson (1990) 'Motives for flight: the Vietnamese refugee exodus, April 1975–1989', Unpublished MA Thesis, University of Hull (April 1990).

he resettlement of people to make way for major infrastructural ›rojects is often justified in developmental terms, it is seldom the elocatee who is the prime beneficiary of such schemes. During the 960s and 1970s, a number of resettlement programmes in northern and orth-eastern Thailand were deemed necessary in order to provide rrigation for rain-fed rice-growing areas and to underpin the country's conomic growth by increasing electricity-generation capacity. Several ens of thousands of farmers were displaced by these schemes, and elocated in resettlement communes, many of which were located in pland areas because of the shortage of land in the lowlands. Whilst

Table 3.2 Numbers of evacuees from selected major dam/reservoir schemes in Third World countries

Country	*Dam*	*Year dam closed*	*Number of evacuees (Nearest '000)*
India	Damodar Valley (4 projects)	1953–59	93,000
Zambia/Rhodesia	Kariba	1958	34,000/22,000
Egypt/Sudan	Aswan	1964	70,000/48,000
Ghana	Volta	1964	82,000
Pakistan	Mangla	1967	90,000
Nigeria	Kainji	1968	44,000
Ivory Coast	Kossou	1971	75,000
Philippines	Upper Pampanga	1973	14,000
Pakistan	Tarbela	1974	86,000
Thailand	(11 projects)	1963–77	130,000
India	Narmada Sardar Sarovar (Gujarat)		70,000
Argentina/Paraguay	Yacyreta		45,000
Brazil	Sobradinho		60,000
Togo	Nangbeto		10,000
China	Shuikou	1987	63,000
China	Danjiangkou	1976	383,000

Source: R. P. Lightfoot (1978) 'The costs of resettling reservoir evacuees in north-east Thailand', *The Journal of Tropical Geography*, 47, 63–74; additional data from Michael M. Cernea (1988) *Involuntary Resettlement in Development Projects: Policy Guidelines in World Bank-Financed Projects*, Washington, DC: World Bank Technical Paper No. 80.

many people have benefited from the irrigation and energy supplies which these schemes provided, the displaced farmers have experienced great difficulty in adapting to their new ecological and economic surroundings.

In contrast to the case of refugees, however, it is generally accepted that it is the government's responsibility to do what it can to assist the reservoir evacuee in adapting to life in a new location. Indeed, the World Bank – a major funding agency for these infrastructural projects – now insists that adequate provision must be made for evacuees before funding is provided. However, because of the vast cost involved in resettling evacuees and short-falls in government funding, resettlement has often resulted in a significant deterioration in living standards for those involved. In part, this is symptomatic of the tendency for evacuees to be seen as a problem to be dealt with as quickly and cheaply as possible, with no real concern for their longer-term welfare. Seldom have relocatees become materially and socially better off than before their move.

Whereas political refugees may leave behind all their worldly possessions, reservoir evacuees may at least be compensated financially for the loss of their land. Farmers and shifting cultivators affected by the Batang Ai hydroelectric power project in Sarawak, East Malaysia, were given quite large amounts of compensation which helped to overcome their misgivings at having to cede their ancestral land to the project. Some 90 per cent of the people displaced by the Damodar Valley projects in India opted for independent resettlement, using the compensation provided by the government to set themselves up in new locations. In the case of the Volta and Aswan dam projects in Africa, however, quite large numbers of displaced people eventually left the official resettlement areas, providing further evidence of the shortcomings of official resettlement programmes.

Voluntary forms of population movement

The final category of population movements which will be described in this chapter involves forms of migration where the mover decides that his or her interests, be they economic, educational, social or whatever, would be better served by moving away from the home community to another place. Voluntary migration is by far the most prevalent type of population movement in the Third World today, taking a wide variety of forms and occurring in response to a very wide range of circumstances. This section will briefly focus on just one form of voluntary movement: that which takes place across international boundaries (international migration). Another very important form of voluntary movement, that which occurs within a national territory (internal migration), provides the focus for discussion in the last three chapters of this volume.

International migration

The extent of international migration involving the Third World should not be underestimated. There are some 350 million people of African descent living outside Africa, compared with around 540 million inside the continent itself. Similarly, there are an estimated 22 million overseas Chinese, and almost 9 million South Asians living outside the subcontinent. In 1984 the International Labour Organization estimated the total stock of economically active international migrants to be between 19.7 and 21.7 million worldwide.

The majority of international movements involving the peoples of the

Third World take place between neighbouring countries, primarily because of the considerable cost which may be involved in travelling large distances. Perhaps the majority of movements between contiguous countries take place within the Third World, although it is those which focus on First World destinations which tend to be the more visible and controversial, and thus receive the greatest attention. Movements across the 'interface' between the Third and First Worlds (e.g. Central America–USA; North Africa–Europe; Asia–Australia) have gained a particularly high profile during the last decade or so as global economic recession and rising unemployment have brought to an abrupt end an era when international migration was strongly encouraged by many First World countries.

Whilst potential economic gain provides the overriding motivation for a large majority of south–north movements, international migration tends to be very restrictive in terms of the kinds of people who are involved. Because of the finance, education, work skills, linguistic ability, sociocultural adaptability and so on which are generally required for this form of movement, a much greater range of potential migrants (notably the poor, unskilled, poorly educated and poorly connected) tend to be filtered out of international migration streams than is the case with other forms of voluntary population movement. This is particularly the case with emigration from the Third World.

Emigration

Since the early 1950s, the global pattern of immigration has seen a significant change from being dominated by European emigrants to the Western Hemisphere and Australasia to consisting primarily of movements from the Third World to the West. Until the early 1970s, most Western countries actively encouraged immigration from Third World countries (e.g. the Caribbean, Latin America and South Asia) as a way of overcoming labour shortages, particularly in manual and unskilled occupations. Figure 3.6 illustrates the range and scale of movement from just one Third World country, Argentina, where during the 1960s and 1970s emigration was underpinned not only by the pursuit of economic opportunities elsewhere but also by political instability and the persecution of certain groups in society.

Increasing rates of immigration provided a convenient scapegoat however, when levels of unemployment began to rise during the 1970s and considerable political pressure was exerted to restrict the scale of

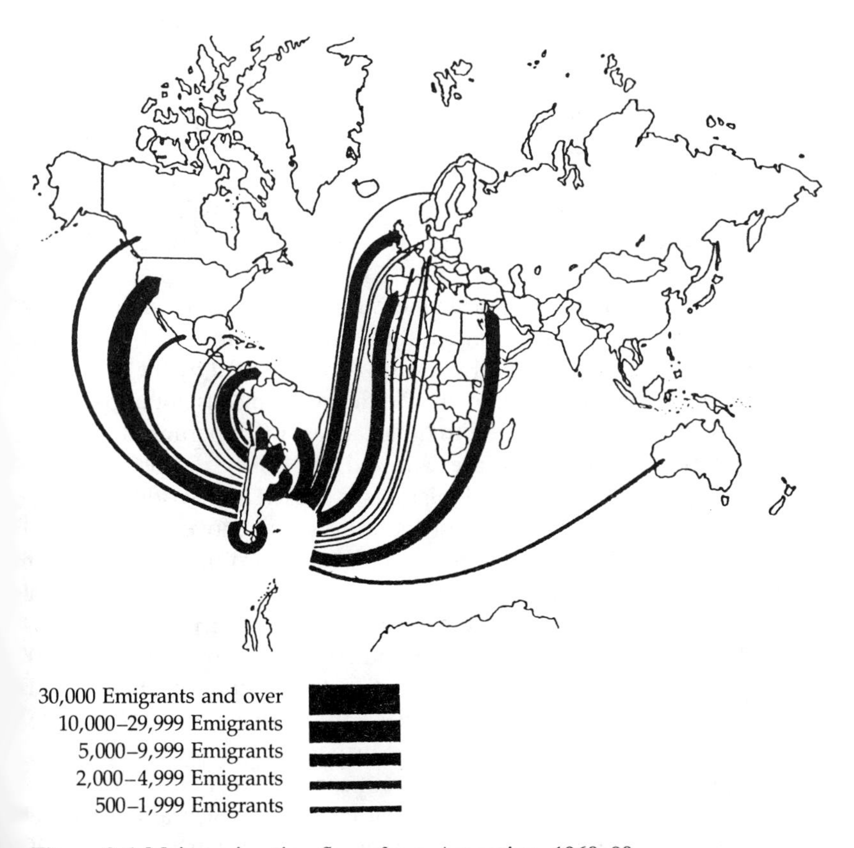

Figure 3.6 Main emigration flows from Argentina, 1960–80
Source: Alfredo E. Lattes and Enrique Oteiza (eds) (1987) *The Dynamics of Argentine Migration (1955–1984): Democracy and the Return of Expatriates*, Geneva: UN Research Institute for Social Development, p. 46.

immigration. Most countries introduced strict quota systems which placed even greater restrictions upon how many people, and which groups in society, could be admitted. Throughout much of the twentieth century the main destinations for Third World emigrants (USA, Canada, Australia and various European countries) have been fairly fastidious about whom they would give immigrant status to. Preference was given to regions with which these countries had been associated historically, or which could supply immigrant workers to fulfil particular economic roles. Following immigration policy reforms in the 1960s and 1970s, discrimination between applicants on the basis of their ethnicity, race, skin colour and nationality has become less prevalent than in the past, but this has been superseded by differentiation on the basis of education, training, skills and other qualities with which applicants might enhance the 'human capital' of the host nation. For this reason, Africa is very poorly represented in the immigration statistics of these major recipient countries. Destitute, poorly educated, unskilled applicants stand little chance of success unless they already have relatives in the host country.

The scale and direction of permanent international migration is thus influenced to a much higher degree than with other forms of voluntary movement by the laws and policies which are enforced by the receiving countries. Unless people operate outside the law, which is not uncommon (see pp. 53–7), they are obliged to obtain visas, work permits and other documentation before they are allowed to travel to their chosen destination. By contrast, internal migrants in most Third World countries are faced with few if any legal restrictions on their movement.

The policies of Third World countries with regard to the emigration of their citizens are quite variable. The East African state of Rwanda is one of a small number of countries which actively encourages emigration as a means of relieving population pressure on scarce land resources. Pakistan also encourages the emigration of skilled and unskilled workers, whilst the Philippines has a fairly ambivalent attitude towards the emigration of its citizens. Furthermore, in spite of concern that the selective nature of emigration constitutes a form of 'brain drain', the governments of a number of countries which already have high rates of emigration (including Egypt, Morocco, Lesotho, Malawi, Haiti and Turkey) are very reluctant to restrict it because of the significant earnings of foreign exchange which are derived from emigrants' remittances. In contrast, a number of Latin American countries, among them Columbia, Ecuador, Guyana and the Dominican Republic, are seeking to limit large-scale emigration amid concern about its effects on domestic development.

Whilst Third World governments in general appear willing to tolerate the emigration of their citizens to the world's more prosperous regions, there is considerable reluctance to assist in the permanent movement of citizens between Third World countries. Very few African states will accept immigrants from neighbouring states for permanent residence. This is partly because they lack the land and financial resources to adequately cope with a large influx of population, but also because they are fearful of the risk of aggravating ethnic or tribal rivalries and conflicts. Immigration in the past of ethnic Chinese and Indians into various South-East Asian countries (e.g. Malaysia, Burma, Vietnam and Indonesia) has created multi-ethnic societies which have occasionally been at the heart of political and racial conflicts.

Illegal migration

Where there are wide differentials in income levels and economic opportunities, considerable flows of population may be expected to result. Where such differentials occur between neighbouring countries, as in the major 'interface' regions between the Third and First Worlds, firm restrictions on the volume of movement between these countries are also likely to be in force. For many, however, the lure of economic opportunities (or their own economic plight) may be such that they may be willing to take the risk of arrest and even imprisonment to avail themselves of these opportunities by illegal means. They may enter a country illegally, perhaps with the help of unscrupulous labour traffickers, and may work without the required documents or permits. Because of their illegal status, they are particularly prone to exploitation by their employers. Their presence may also engender considerable resentment from other citizens, fearful of the effects of illegal workers on their own jobs and rates of pay.

The scale of illegal migration between countries is very difficult to ascertain because of its very nature. Estimates of the number of illegal workers in the United States vary from 2 to 12 million, although a more realistic figure is around 5 million, or some 2 per cent of the country's population. Perhaps as many as half of these are Mexicans (see Case study B), together with sizeable numbers of Cubans, Haitians, Dominicans and citizens of other neighbouring countries. It has been estimated that up to half a million illegal residents are being added to the population of the United States each year. The scale of illegal migration is probably increasing, due to widening wealth differentials, improvements in transportation and increasing restrictions on legitimate forms of immigration. Figure 3.7 shows that the scale of illegal migration in the United States, as evidenced by the number of illegal aliens who were taken into custody, increased quite sharply with the termination in 1964 of the Bracero Program which had allowed Mexican migrants to enter south-western states as agricultural workers.

The phenomenon of illegal international migration is not restricted to

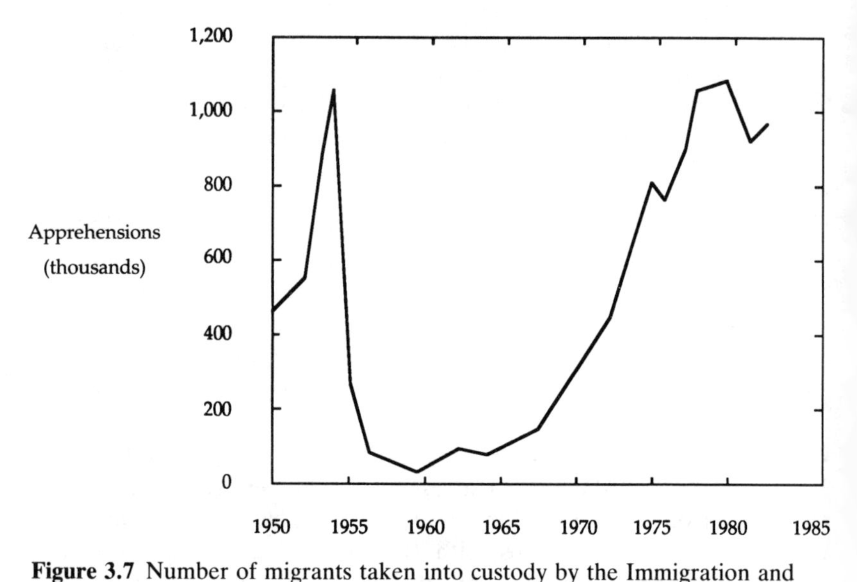

Figure 3.7 Number of migrants taken into custody by the Immigration and Naturalization Service in the USA, 1950–80
Source: Charles B. Keely (1982) 'Illegal migration', *Scientific American* 246 (3), p. 34.

the United States, of course, nor does it consist exclusively of movements from the Third World to the First. Tens of thousands of mainland Chinese attempt to enter Hong Kong illegally each year. Within South-East Asia there is a considerable volume of illegal population movement between individual countries, reflecting economic differentials within the region. Thus there are several thousands of illegal migrants from Thailand and Malaysia currently working in Singapore, and possibly as many as 250,000 Indonesians living illegally in Malaysia. In Latin America there is also a significant volume of illegal movement towards the main centres of economic opportunity in Venezuela (up to 2 million) and Argentina (around 1 million) and elsewhere. Some 60,000 illegal Haitian migrants were expelled from the Bahamas in the late 1970s. In Africa, the oil-rich state of Libya provides an important focus for illegal migrants, particularly from neighbouring Tunisia and Egypt.

International labour migration

Whilst the tendency over the last decade or so has been for greater legislative barriers to be erected against international migration, there

Case study B

Illegal migration from Mexico to the USA: across the river and into the States

From the door of her apartment at number 812 on Oregon Street, Maria Conception Robles de Villegas, or Concha, as she is known to her family, can see two towns, a muddy river and the bridge that links the two worlds of her schizophrenic existence. She can also see the street corner where her son was nearly born 10 years ago. Concha was already in labour when she crossed Santa Fe bridge from Ciudad Juarez in Mexico, determined that her son would be born in the US. 'He nearly came out on the bridge. What a problem, he would have been stateless,' says Concha. 'But in the end he was born American.'

Her family is scattered on both sides of the Rio Grande. Her 10-year-old son goes to the state school in El Paso, but spends his weekends with his cousins in Juarez. 'My loved ones live on both sides, and I'm split between the two,' Concha says. 'My blood calls me from the other side, but I think it's best for my son to get his education here, in his own country.'

Concha Robles de Villegas is a typical inhabitant of the frontier land that lies between the US and Mexico. It is a world of shared loyalties, divided families, economic self-sufficiency, and mixed and blurred customs, cultures and laws. Carmen and Rodriga have just crossed the Rio Grande illegally from Mexico. They are going shopping at J. C. Penney's, the American store. They say it sells cheaper, better-quality trainers than those sold in their home town.

El Paso is where the First World meets the Third. Only a small river divides them.

Javier Hernandez has just sold the last of the fruit he brought from Juarez at midday, crossing the Rio Grande illegally as he has done twice daily for four years. He has been stopped about 20 times by immigration officers. They take him to a detention centre, fill in a prosecution form, and 15 minutes later take him back to the south side of the bridge. Shortly afterwards, he crosses the river again to pick up his bicycle. 'There isn't any work in Mexico,' says Hernandez. He is not exaggerating. Unemployment

Case study B *(continued)*

in Mexico runs at 40 per cent. Another 20 per cent of adults are estimated to be underemployed. Mexico's population stands at about 85 million, of which 40 million are under 18 years of age. In El Paso, 69 per cent of the inhabitants are of Mexican origin.

On the Mexican side of Santa Fe bridge, Benito Juarez Avenue takes over from El Paso High Street. It is full of things that the neighbours from El Paso come over to buy. 'Painless extractions' reads the sign in a dentist's window. They are very popular with the neighbours from El Paso as they are so much cheaper. A few feet further on are the discotheques where you can buy two dances for a dollar. In a side street is the Jose Lopez Vega Maternity Clinic, just beside Hotel Lido, which charges $7 an hour.

At 8 a.m. in the general headquarters of the El Paso sector, officer Jimmy Walker of the Border Patrol is drinking his first coffee of the day. A veteran of 27 years in the patrol, he acknowledges that 'I've worn a revolver for more than half my life'. This morning he is going to scout along the frontier, from the fence the Mexicans call 'the tortilla fence' because it is so easy to break through, to the desert zone under the jurisdiction of the mounted patrol, behind the mountain of Cristo Rey.

The patrol's 640 men cover the 520 kilometres of El Paso sector, a territory of some 205,000 square kilometres. 'We don't take this work too seriously,' says Walker. 'We do the best we can, given the resources and bearing in mind that we want to maintain friendly relations between Mexico and the US.' In 1990 the patrol arrested more than 1 million people who crossed the frontier illegally and returned them to Mexico. In the El Paso sector, the corresponding number was 223,189. Doug Mosier, patrol spokesman, says: 'We know that if we catch 223,000, at least another 400,000 got through.' Of course, less than 10 per cent come as permanent immigrants. Most cross for the day, the week, or a season of work. In a survey published last March, the Rand Corporation of California estimated that some 135,000 Mexicans emigrate illegally every year. There are 12.4 million Mexicans living in the US.

It is easy to see why so many cross the river undetected. In the

Case study B *(continued)*

places where the river is shallow, people simply take off their shoes and walk across. In others they can cross by jumping from stone to stone. In many of the deeper spots, there is an informal transport system of boats made from truck tyres. The charge is a dollar a trip. When the Mexicans see Walker, they just smile.

Source: Roberto Fabricio 'Across the river and into the States', © The *Guardian*, 14 June 1991, p. 26.

are a number of countries which, because they have labour needs that cannot easily be satisfied from their domestic workforces, have actively encouraged in-migration. Western European countries until the oil crises of the 1970s and, paradoxically, Middle Eastern states which since the 1970s have been awash with 'petrodollars', have encouraged foreign workers, mostly from poorer semi-periphery and Third World countries, to engage in various forms of labour migration to these labour-scarce regions. Disparities in the level and pace of development between Third World countries have also encouraged a considerable volume of international migration of workers within the Third World. This section will conclude with a brief look at these various forms of international labour migration.

Third World–First World

Western Europe has played host to a large volume of foreign migrant workers, particularly since the period of reconstruction and rapid economic development that followed the Second World War. Initially, the majority of workers came as temporary migrants from the poorer regions of Mediterranean Europe (particularly Italy, and later Spain, Portugal and Greece), and were brought to countries such as France, West Germany, the Netherlands and Belgium via bilateral labour recruitment agreements. The rising pace of economic growth, and heightening competition for foreign workers caused several Western European countries to look further afield to satisfy their labour needs. A number of countries opened the doors to large-scale immigration from their former colonial territories (e.g. from South Asia and the Caribbean to Britain, and from North and West Africa to France), as

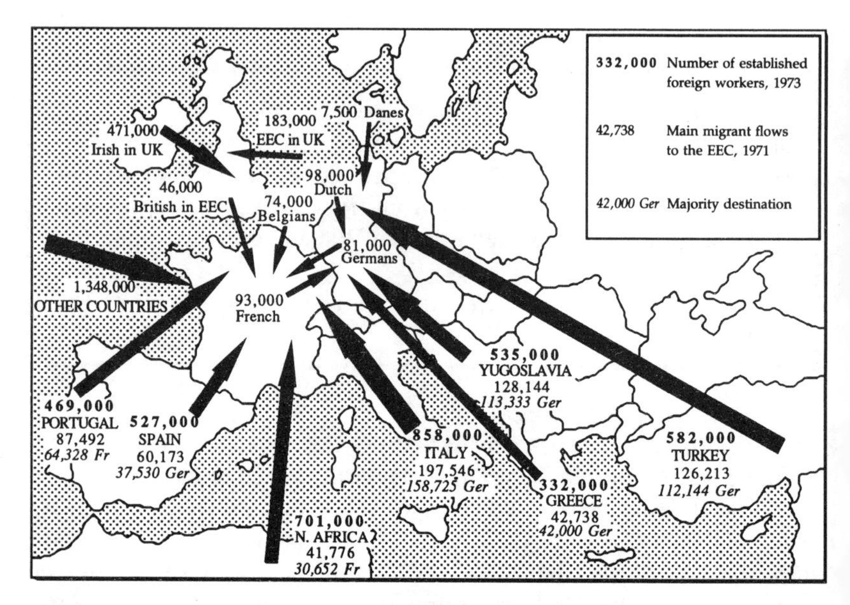

Figure 3.8 Patterns of labour migration into selected European countries in 1971

Source: G. J. Lewis (1982) *Human Migration: A Geographical Perspective*, London: Croom Helm, p. 157.

well as casting their recruitment nets beyond Europe – to countries lying at the 'interface' between the First and Third Worlds (e.g. Turkey, Algeria, Morocco, Tunisia).

Migrant workers from the latter countries were invited as 'guest workers' who, it was envisaged, would remain in the host country for only a matter of years before returning to their home countries. Partly because of their temporary status, host governments and employers were reluctant to invest heavily in social infrastructure for their 'guest workers'. Accordingly, most had to put up with very poor living conditions and enjoyed few of the welfare benefits which their host societies took for granted. In addition, because the majority of migrant workers were concerned to save as much as possible from their earnings to send home to the families they had left behind, they were themselves reluctant to invest scarce funds in improving their living conditions. Although government legislation protected the rights of migrant labourers recruited through formal bilateral agreements, the large volume of spontaneous and clandestine migration to Western Europe at this time made it extremely difficult for governments to exercise complete control over the way that foreign workers were treated. Most 'guest workers' filled poorly paid unskilled and semi-skilled occupations in industry, construction and the service sector.

The number of foreign migrant workers in Western Europe rose steadily until the early 1970s. Thereafter the volume of movement stabilized, and formal recruitment programmes ceased. Following the passing of EEC legislation in November 1973, severe restrictions were placed on the movement of non-EEC workers into the member states. The effect of these restrictions in the case of migration into West Germany is strikingly evident from Figure 3.9. Furthermore, and as a result of global recession, levels of unemployment began to rise rapidly. Foreign workers were the first to be affected, partly because they filled many of the more marginal and dispensable occupations, but also because of growing political pressure which tended to scapegoat foreign workers as being responsible for rising unemployment. Right wing political parties jumped on the anti-immigration bandwagon, and in many European countries there was a strong backlash against immigrant communities. In 1973 Algeria suspended all emigration to France because of attacks on Algerians in France: 32 Algerians were murdered in France in the same year.

Much as governments would have liked their 'guest workers' to return home, many were reluctant to do so. Most would have faced great difficulty in finding employment in countries where their departure had helped to alleviate domestic unemployment levels. These countries were also not immune to the global economic problems which were affecting the countries of Western Europe. Furthermore, many of the migrant workers had, over time, been joined by their families and had become settled in their host countries. Many had children who had been born in the host country.

Very reluctantly, the governments of France, West Germany and the Netherlands, amongst others, accepted that the immigrant worker problem would not disappear of its own accord, and they instituted a series of programmes targeted at the foreign population. The West German government, for example, introduced a series of incentives aimed at encouraging Turkish migrant workers to return home, including paid passage and a resettlement grant. At the same time it has gradually narrowed the criteria which determined the eligibility of people to enter the country for family reunification. The Dutch government, on the other hand, has accepted a large proportion of its foreign

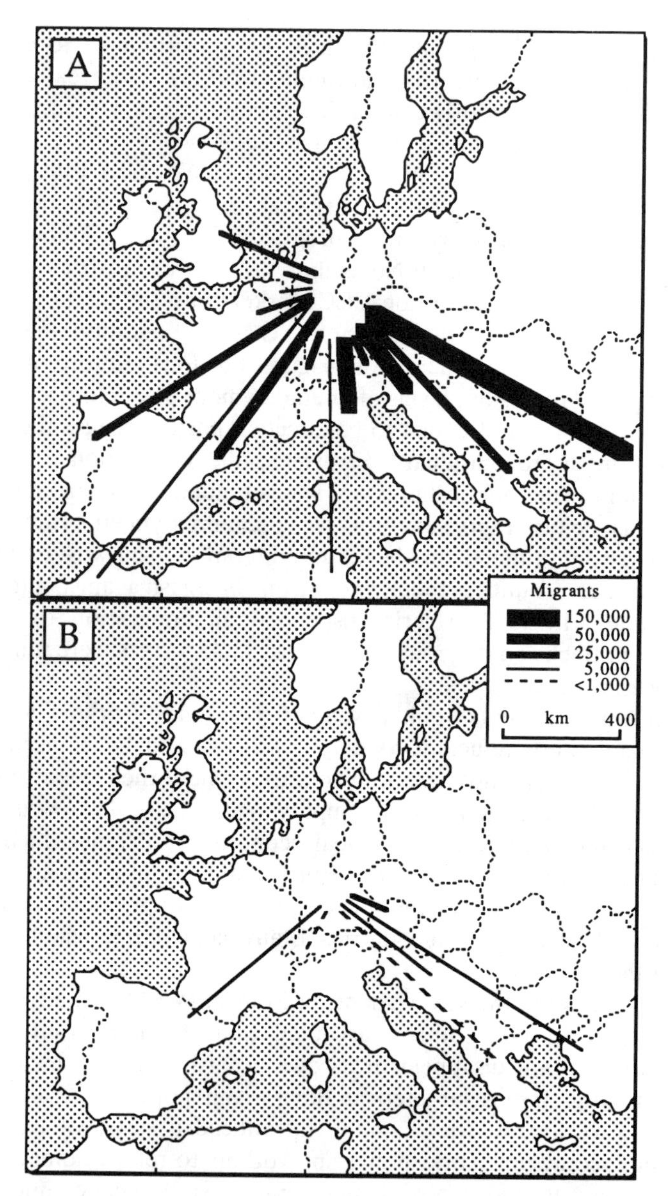

Figure 3.9 Migration of foreign workers into West Germany in 1972 (A) and 1981 (B)
Source: John Salt (1985) 'West German dilemma: little Turks or young Germans?', *Geography* 70 (2), p. 164.

population as permanent residents, and introduced a series of policies aimed at promoting their full participation in society as 'ethnic minorities'.

Third World–Middle East

The pattern of movement of migrant workers from Third World countries to the Middle East displays a similar picture of a boom period followed by a steady decline in demand for foreign labour. Until the dramatic exodus at the time of the Gulf Crisis, there were in excess of 3 million Asian labour migrants (and almost as many African foreign workers) working in the Middle East. At the beginning of the 1970s there had been only about 200,000. This dramatic increase was explained in part by the immense wealth of oil-producing countries in the Middle East following the oil price hikes of 1973/4 and 1978/9, and in part by the growing employment crises of several poorer Asian countries.

Table 3.3 Stock of Asian workers in the Middle East in the early 1980s for the main sending countries (estimated)

	Year	*Total number of labour migrants*
Bangladesh	1982	240,000
India	1983	800,000
Pakistan	1983	1,220,000
Philippines	1983	710,000
South Korea	1982	173,000
Thailand	1982	200,000

Source: Derived from United Nations. (1987) *International Labour Migration and Remittances Between the Developing ESCAP Countries and the Middle East: Trends, Issues and Policies*, UN, Economic and Social Commission for Asia and the Pacific, Development Papers, no. 6, p. 17.

The oil-producing states (particularly Saudi Arabia, the United Arab Emirates and Kuwait) embarked on major investment programmes aimed at improving their social and economic infrastructure. However, their domestic workforces were totally inadequate for this task. Populations were very small, and were mostly located in rural areas. The female participation rate in the workforce was very limited, and levels of education and training were generally low. There was also a general reluctance among the workforce to become engaged in manual work. The response was to import labour, initially from labour-surplus countries elsewhere in the Middle East such as Egypt, North and South Yemen and Jordan, and later from South and South-East Asia (see Table 3.3). Here, the pace of economic development was generally very slow, domestic employment opportunities limited and income levels depressed. In contrast, employment on the various construction projects in the Middle East offered the prospect of, by domestic standards, very high wages. Asian workers were thus very keen and quick to exploit these opportunities. Their respective governments were equally keen to encourage large numbers of migrants to move to the Middle East, seeing the movement as an 'export of labour' which would serve the dual purpose of earning precious foreign exchange and alleviating problems of unemployment at home.

Third World–Third World

Inequalities in the level and pace of development within the Third World have also resulted in the movement of workers from poorer to richer Third World countries. The majority of such movements naturally take place between neighbouring countries and over relatively short distances. The influence of uneven development on patterns of labour migration is evident throughout the Developing World. In South-East Asia, the tiny oil-rich Sultanate of Brunei has become a major focus for

Plate 3.2 Living quarters for Thai construction workers in Singapore. Several thousand Thai men are employed in Singapore's massive public housing and transportation programme

migrant workers from Thailand, the Philippines and the neighbouring states of East Malaysia, where the men are engaged as construction and oil industry workers, and the women as *amah* (housemaids). In the wealthy island state of Singapore, migrant workers from Thailand and the Philippines, many of them illegal, have provided the labour resources for the Republic's massive public housing and transportation programmes. Between 15 and 20 per cent of Singapore's labour force is made up of foreign workers. It is ironic that Thailand, with one of the world's most rapidly expanding economies, continues to export large volumes of semi-skilled and unskilled labour. This too is symptomatic of the fact that the spoils of Thailand's recent economic growth have not been spread evenly, leaving a large residual of poor people who are anxious to exploit economic opportunities wherever they may be found.

Oil wealth and rapid economic development also underpin international labour migration in other parts of the Third World. Venezuela has provided a focus for labour migration, both legal and illegal, from Columbia, Ecuador and several Caribbean states. Libya is a major labour-importing country in North Africa. Elsewhere in the Third World international labour migration is influenced by the relative distribution of population and economic opportunities, and by state policies towards the movement of people across national frontiers. Several African countries encourage the export of labour as a way of alleviating unemployment and boosting foreign exchange earnings. Morocco has sought to fill the vacuum caused by restricted access to European labour markets by encouraging labour migration to neighbouring countries, although the scope for this is limited by the restricted range of marketable skills offered by the Moroccan workforce. In western Africa, the Ivory Coast has for a long time been a major recipient of temporary migrant workers from Mali, Togo and Upper Volta who have migrated to work on the plantations and in other fields of employment which are shunned by nationals. In the late 1970s, the country's 1.4 million foreign workers made up 21 per cent of the total population and 35 per cent of the workforce. Reflecting the way that economic fortunes may change, Ghana, once a major destination for migrants from elsewhere in West Africa, became a source of large-scale emigration during the 1980s as a result of a sharp economic down-turn. A major stream of migration developed between Ghana and Nigeria until, with declining oil prices in the 1980s, large numbers of Ghanaian workers were expelled from the country.

Case study C

International and internal labour migration in southern Africa

The resource wealth and relative economic dynamism of South Africa has supported a system of migration which involves not only large numbers of workers from within the Republic but also international migrants from neighbouring southern African states. Labour migration to the mines, farms and factories of South Africa is unique in the Third World setting because of the social and political system within which it operates. The system of apartheid has excluded the majority of black workers from the cities and residential areas of white South Africans, and has forced their concentration in the poor, deprived and unsanitary surroundings of the townships. Labour policies force male migrants from further afield to endure weeks or months of separation from their families. Black workers are seen and used largely as anonymous 'units of labour', with little attention being paid to their dignity and rights as citizens and human beings. Although recently the situation in South Africa has been changing quite rapidly, the legacy of apartheid and the institutionalized subordination of black workers will continue to affect the lives of several million people for quite some time to come. Neighbouring countries also face an uncertain future as opportunities for migration – an important source of foreign exchange – may be expected to decline steadily during the 1990s. The following short case study examines some of the more important features of internal and international migration in southern Africa.

Internal migration

The origins of a migrant labour system in South Africa can be traced to the discovery of diamonds in Kimberley in 1867, and gold in the Witwatersrand in 1886. Large numbers of black workers moved to the towns and cities which sprang up in the mining areas, attracted by opportunities for wage employment. The influx led to demands from whites for the segregation of the two main races. This was achieved via the 1923 Natives (Urban Areas) Act, which enforced strict controls over the influx of black workers into the

Case study C *(continued)*

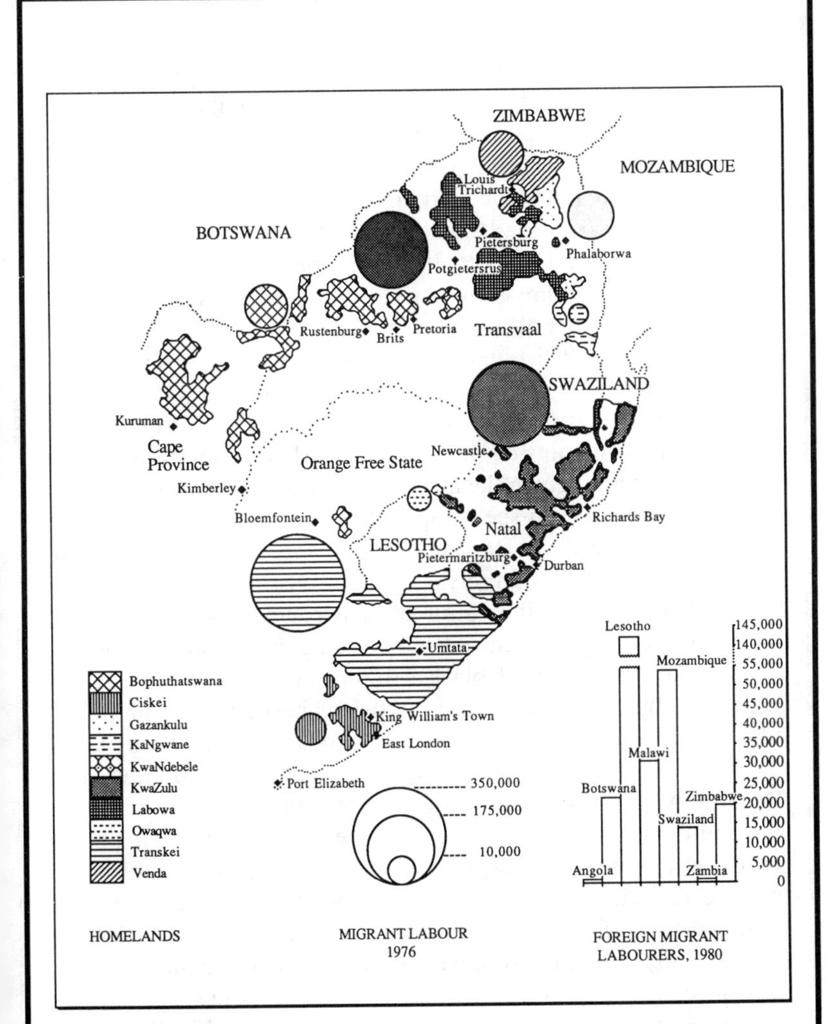

Figure C.1 Migrant labour in South Africa, indicating homeland and foreign sources

Source: Anthony Lemon (1983) 'Migrant labour and frontier commuters: reorganizing South Africa's black labour supply', in David M. Smith (ed.), *Living Under Apartheid: Aspects of Urbanization and Social Change in South Africa*, London: Allen & Unwin, p. 70.

Case study C *(continued)*

cities and which established the framework for the system of repression and control which is only today being gradually dismantled. The 1923 Act facilitated the manipulation of black migrant labour through a controlled system of labour supply; it led to the establishment of segregated townships; it greatly restricted the political rights of black workers in urban areas; and it imposed strict constraints on land ownership, which effectively excluded blacks from the land market. It also established the principle that migration to urban areas by black workers would be temporary: there were no rights of permanent residence.

The 1913 and 1936 Land Acts set aside 'reserves', bantustans, mostly in marginal and peripheral zones, for the settlement of black South Africans, and meanwhile prohibited them from acquiring or leasing land outside these reserves. Most struggled to subsist in these areas and, given the added burden of heavy taxation, were more or less forced to seek wage labour in the mines and on the farms of white settlers. Deprived of male workers, the economies of the bantustans have largely stagnated. Overcrowding has led to severe problems of soil erosion and environmental deterioration, with drought and malnutrition becoming increasing problems. Meanwhile, the influx of black workers to the cities was strictly controlled by various 'pass laws'. Workers needed to obtain passes to enable them to gain access to urban jobs. The pass laws played an important role in reducing the cost of labour in South Africa. Because workers retained a subsistence foothold in the reserves, the South African authorities and businesses felt justified in paying very low wages, providing the barest minimum of amenities and welfare facilities, and allowing workers no local political rights. Wages were sufficient only to support an individual rather than an entire family, and laws were passed which prohibited the provision of married accommodation for migrant workers. Thus the majority of household members remained in the home area. Families in the reserves shouldered a heavy burden in supporting migrant workers, especially during periods of illness or unemployment.

The exclusion of black workers from the cities, and their concentration in the bantustans, has served to separate, sometimes

Case study C *(continued)*

by a considerable distance, the places where people work and live. This has given rise to several, simultaneous patterns of migration in South Africa. Large numbers of people commute daily or weekly from the bantustans, and the townships which have sprung up, mostly illegally, nearby the major urban centres. Commuters spend on average between 2.5 and 4.5 hours each day travelling between home and work. People living too far away to be able to commute will live, separated from their families, in single-sex urban hostels or townships while they see out their work contracts. Surplus labour and squatters are forcibly removed from the cities and townships: since 1960 more than 3.5 million people have been uprooted in this way. Since 1968, all legal workers from the bantustans have been considered contract labourers and must return to their homelands every 11 months from where they can be re-employed.

Some quite fundamental changes have taken place over the last five years or so in policies affecting internal migration, although they have done little to change the *status quo*. In 1986 the South African government abolished its influx control laws, including the Group Areas Act, principally because they were increasingly being ignored, both by migrant workers and urban employers. They were replaced by a policy of 'orderly urbanization', in which black workers were theoretically allowed to move into the cities. In reality, the urban influx continued to be restricted by the government's control over their access to 'approved' accommodation and land for housing, and the enforcement of strict measures to control squatting. Also, few black South Africans can afford to purchase property in white residential areas. The government of President F. W. de Klerk has also repealed the Land Acts, in theory enabling black Africans to own land outside the bantustans. This means little in practice, because most blacks are too poor to be able to afford to purchase land. An important nail in the coffin of apartheid was struck on 17 June 1991 when South Africa's Parliament voted to repeal the Population Registration Act, under which all citizens are classified according to their race.

Case study C *(continued)*

International migration

Since 1973, South Africa has steadily reduced the amount of foreign labour which it employs – which in turn has led to some severe employment problems in source countries (especially Lesotho, Swaziland and Botswana, and to a lesser extent Mozambique and Malawi). Up until the mid-1970s these countries had come to rely very heavily upon international migration not only for the foreign exchange provided by migrants' remittances but also as a much-needed employment outlet for a workforce facing severe economic stagnation on the domestic front. So great was their dependence upon labour migration to South Africa that they were unable to exert any real pressure on the South African government to improve living and working conditions, wages and rights for the migrant workers. At the same time, the South African government had used labour demand as a powerful lever for punishing dissent and for prising political support from neighbouring territories, particularly Lesotho which in the mid-1980s relied on labour migration for more than half its Gross National Product. As a result, while the demand for foreign labour in South Africa virtually halved between 1973 and 1980 in favour of domestic labour, Lesotho was virtually unaffected.

Things began to change during the mid-1970s. Malawi withdrew some 120,000 workers from South Africa's gold mines in 1974 following the Francistown air crash in which 82 labour recruits from Malawi died. Mozambique also withdrew foreign workers as a political gesture and as a way of reducing its reliance on the Republic of South Africa. The labour deficit which resulted from the withdrawal of these two important source countries was offset by increased recruitment of contract labour from Lesotho. More recently, the Republic has pursued an 'internalization programme' wherein it seeks to substitute domestic labour for foreign workers. This programme has been helped by a rise in the price of gold and a subsequent rise in wage rates which has increased the attractiveness of mining *vis-à-vis* industrial employment. As a result, the proportion of foreign black workers in the mining labour force fell from around four-fifths in the early 1970s to two-thirds in the mid-1980s.

Case study C *(continued)*

South Africa could potentially satisfy all its future labour targets from the domestic workforce, which leaves Lesotho, Botswana and Swaziland in a very vulnerable position, at least until they have created more employment opportunities in the domestic economy. In the meantime, they are in a very weak position in terms of being able to exert pressure on the South African government to improve workers' rights and conditions during their contracted periods of employment in the Republic.

Sources: David Simon (1989) 'Rural–urban interaction and development in southern Africa: the implications of reduced labour migration', in Robert B. Potter and Tim Unwin (eds), *The Geography of Urban–Rural Interaction in Developing Countries: Essays for Alan B. Mountjoy*, London: Routledge, pp. 141–68; Dhiru V. Soni and Brij Maharaj (1991) 'Emerging urban forms in rural South Africa', *Antipode* 23 (1), pp. 47–67.

Conclusion

This chapter has highlighted the great variety of forms of population movement which occur in the Third World today. For every type of movement which has been discussed there are several others which constraints of space have prevented from being included. It should be clear from the foregoing discussion that the process of development, which includes social and political as well as economic changes, plays a quite fundamental role in the initiation and perpetuation of migration flows between and within countries and regions of the world. The following chapter will examine in more detail how the development process influences the mobility decisions of people in the Third World, using internal migration as the main subject for discussion. Chapter 5 will then assess how the process of migration influences the development of rural and urban areas in Third World countries.

Key ideas

1 Long-established traditions of mobility which are deeply rooted in the cultures of many Third World societies may have paved the way for contemporary forms of economic migration.
2 Demographic and commercial pressures have made it increasingly difficult for people to undertake periodic movements to enable them to adapt to living in marginal ecological zones.
3 As governments have imposed ever more stringent controls on immigration, so the pressure and incentive for clandestine or illegal international migration has increased.
4 The distinction between a 'political refugee' and an 'economic opportunist' is sometimes very difficult to make.
5 In areas with some potential for major infrastructural schemes, the wider national interest almost always prevails over the interests of the people who will be displaced by their construction.

4
Why people move

> Migration takes place when an individual decides that it is preferable to move rather than to stay and where the difficulties of moving seem to be more than offset by the expected rewards (L. A. Kosinski and R. M. Prothero (eds) (1975) *People on the Move: Studies on Internal Migration*, London: Methuen).

This statement encapsulates a number of important features of the migration process in Third World countries. It indicates that movements generally take place in response to the circumstances, actual as well as potential and perceived, with which people are faced both in their home communities and in areas away from home. The problems, opportunities and changes associated with the development process provide the main motivation for movement in most Third World regions. The statement also shows that migration is usually preceded by a process of decision-making in which the advantages and disadvantages are carefully weighed up, and where the potential difficulties associated with migration may be traded off against those which might result from staying put.

The statement none the less presents an incomplete and rather too simplistic view of the migration decision-making process. This is only to be expected, given the great variety and complexity of factors which underpin the decision to move. The suggestion that migration represents an entirely rational course of action which is taken in response to a reasoned and well-informed judgement about conditions elsewhere is questionable. Very often people migrate more in hope than expectation of finding a better life elsewhere. Some simply end up moving from one environment of poverty and exploitation to another.

The image of a decision-making process involving individuals is also an inappropriate one in a great many cases. Migration decisions are seldom taken without due consideration of the broader implications for the migrant's family, household or community at large (the 'dependent' category of movements which were outlined in Chapter 2). Indeed, sometimes the migrant is the reluctant party to decisions which are taken by someone with a higher authority (typically a parent or village elder). This helps to explain why many migrants experience great difficulty in coming to terms with their new circumstances, and very quickly yearn to return to the more familiar and less intimidating surroundings of the home community.

The suggestion that migration is a relatively predictable and homogeneous form of action must also be contested. If conditions are so poor in the source area, why is it that only a fraction of households in that area (admittedly quite a large fraction in some cases) will respond by sending someone to search out opportunities elsewhere, whilst others appear content to trust their luck to remaining *in situ*? Clearly, whatever factors influence the migration decision-making process, they do not affect everyone to the same extent. Whilst there is a need to understand the general reasons why people move from one place to another, not least from the viewpoint of development planning, we should none the less also be mindful of the huge complexity and heterogeneity of the migration process. Migration occurs in response to a wide range of factors which affect different people in different ways, and to which people do not necessarily respond in an identical fashion.

This chapter aims to identify some of the key factors which help to explain why people move, and to show how differences in the pattern and process of development in Third World countries influence the scale and character of population movements. The reasons why people move are examined at three overlapping levels of enquiry. At the micro-level we look at some of the factors which compel individual migrants to leave their home areas, and at how and why the migration decisions of individual households are made. At the meso-level, we seek to explain patterned regularities in the migration process in terms of prevailing social and economic conditions in major source and destination areas. We start by examining the influence on migration of a variety of macro-level processes as changes which are influencing the pattern, pace and process of development within the Third World. The discussion focuses on the influence on migration of development at these different levels: readers are referred to several of the books in this series for a more detailed examination of the characteristics of development in Third World countries.

The macro-level perspective

> [W]e can never specify the exact set of factors which impels or prohibits migration for a given person, we can in general, only set forth a few which seem of special importance and note the general and actual reaction of a given group.
>
> (E. S. Lee (1966) 'A theory of migration', *Demography* 3: 47–57.)

Inequalities in the pattern and process of development provide the back-drop against which a great deal of migration involving the Third World can be viewed. We saw in Chapter 3 how economic differentials between nations underpinned a great deal of the international movement of population which originates in Third World countries. Migration provides a mechanism whereby the 'human resources' of countries where labour is abundant and underutilized can be productively employed in satisfying the labour needs of countries where development is being constrained by domestic labour shortages, particularly for manual and semi-skilled tasks. The existence of wealth disparities between nations does not in itself *inevitably* result in large-scale movements of economic migrants: much depends upon the ability of source countries to supply, and host countries to absorb, migrant labour, and upon state policies with regard to the freedom of movement of workers across international boundaries.

Our first clues as to the influence of uneven development on migration *within* Third World countries are provided by the prevailing and predominant patterns of movement therein. We identified in Chapter 2 a tendency for the majority of movements to take place from the countryside towards the city, and from economically depressed peripheral regions towards the main centres of economic activity. Why should this be the case? The explanation lies mainly in the structure of economic development in these countries, and in particular the widespread tendency towards the concentration of economic growth and activity in key dynamic areas and sectors. This may partly be the result of the natural tendencies of free-market capitalism, differential endowments of natural resources, the strategic geographical importance of some areas, or deliberate government policies which afford a higher level of investment and support to more prosperous and dynamic regions. Away from these key areas, economic development tends to be much slower, as does the pace of technological change and innovation. Migration may thus be seen as people 'voting with their feet' by moving away from depressed regions and seeking to gain access to the opportunities and rewards which they perceive to be available in the core regions. In this way migration can be viewed as a response to the unevenness of the development process.

Several of the volumes in this series have examined the structural imbalances which have been associated with Third World development, and these too have had an important impact on the incidence and characteristics of migration. In short, these imbalances are largely the consequence of the development strategies which have been employed in these countries over the last century or so. The period of colonial domination led to the selective and incomplete opening-up of the territories in Third World regions, and supported development in a restricted range of economic sectors. Associated with this process of unbalanced development, the migration of labour was often encouraged to facilitate the construction of various forms of infrastructure and also to provide the workers for colonial enterprises.

Capitalism arrived on the coat-tails of colonialism, and its introduction into non-capitalist societies had (and continues to have) a profound impact on the incidence of migration. The labour needs of capitalist production, such as in mining and plantation agriculture, were satisfied by restricting peasants' access to land resources and by coercing people (directly through forced labour systems, and indirectly through taxation) into migration to work as waged labourers in the capitalist sector. The communications systems which developed in support of capitalist enterprise helped to facilitate the mobility of human resources. As capitalism spread into peripheral and more isolated areas, so the imperatives of migration to satisfy growing cash needs have also expanded. Thus capitalism simultaneously created a demand for migrant labour and a set of imperatives which encouraged people to seek employment and income via migration. The spatial development and impact of capitalism has been very uneven, and this continues to be reflected in the pattern of labour movements within Third World countries.

Since the Second World War, almost all countries of the Third World have sought to bring about development by facilitating rapid economic growth, principally through the intensified exploitation of natural resources, the commercialization of agriculture, and industrialization. This last process has had a particularly powerful impact on the level and pattern of migration, with people attracted to urban areas in increasing numbers by the prospect of employment and higher wages in industry. The encouragement given to the urban-industrial sector stands in stark contrast to the benign neglect that rural areas have had to endure on account of their limited capacity to underpin rapid economic growth. The stagnation and relative backwardness of the rural-agricultural sector (often stronger in perception than reality) provides further impetus for the movement of population in Third World countries, and yields further evidence of the importance of macro-level economic disparities in underpinning migration in Third World regions.

However, a shortcoming of the macro-level structural perspective is that it tends to view migrants as an amorphous, homogeneous entity who appear to have little effective choice but to migrate. In reality, as we have seen, perhaps the majority of people respond in a neutral manner to differentials in the pattern of development. The macro-level approach offers few explanations as to why some people, ostensibly faced with the same circumstances, prefer to stay put, and also why not all forms of migration occur in the direction of the economically more dynamic regions. Thus the simple existence of structural imbalances does not, by itself, provide sufficient reason for migration to occur. For this to happen, these imbalances have to impinge directly on people's daily lives, and people must want to avail themselves of the opportunities and rewards that migration potentially offers in spite of the difficulties and disruptions that this may entail.

Thus, whilst the structural perspective is useful in providing a broad framework for understanding the incidence of migration in relation to the development process, there is clearly also a need to show how these general macro-level processes translate into real-life situations. This is achieved by adding the meso-level dimension to our examination of why people move which, for the sake of simplicity, will be taken to include factors in the places of origin and destination which influence people's migration decisions.

The meso-level perspective

A useful starting point for examining both why people move and why they move in particular directions is provided by the model developed

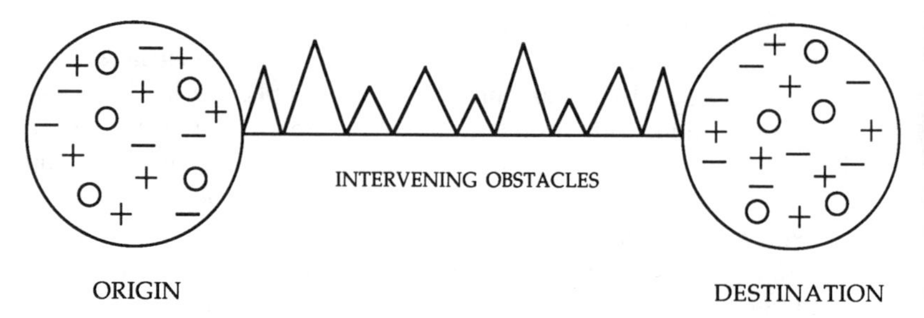

Figure 4.1 Everett Lee's model of origin and destination factors and intervening obstacles in migration
Source: Everett S. Lee (1966) 'A theory of migration', *Demography* 3, p. 50.

by Everett Lee in the mid-1960s (see Figure 4.1). The model, which drew its inspiration from the 'laws of migration' which were suggested by E. G. Ravenstein at the end of the nineteenth century, sought to outline some of the main factors which enter into the decision to migrate. Although it has been criticized for the way in which it simplifies such a complex process, the model none the less helps to highlight the importance of meso-level influences on the migration decision-making process.

Migration takes place, Lee argued, in response to the prevailing set of factors both in the migrant's place of origin and in one or a number of potential destinations. These factors were identified by Lee as being positive (+), negative (−) or neutral (o). In simple terms, migration is seen as being most likely to take place where the influence of negative conditions in the place of origin and/or positive conditions in a potential place of destination is greater than the conditions which attach people to their home areas or dissuade them from moving elsewhere. The relative balance of positive and negative conditions may have a powerful influence not only on the incidence of migration but also upon the direction of movements, with streams of migration developing towards the more attractive destination areas. It may also have an influence on the duration of movements, with return migration being more likely to occur where the home community continues to hold an attraction to the mover. A variant of Lee's model is the 'push–pull' framework which similarly views migration as a response to repulsive forces in the place of origin and attractive forces in the place of destination.

Lee however cautioned against the use of a simple calculus of positive and negative factors to determine the likely volume of migration. He suggested that the factors in favour of migration would generally have to outweigh substantially those against, because of people's natural reluctance to uproot themselves from the familiar surroundings of their home areas. It is also important to note that people cannot be expected to respond in an identical manner to these positive and negative forces. In the final analysis, much depends upon the individual's own circumstances, their personality, and their perception of the conditions which surround them. None the less, as Lee puts it, some factors affect most people in the same way, and other factors affect different people in different ways. These divide quite neatly into the meso-level determinants of migration (outlined below), many of which are associated with the development process, and the micro-level factors which influence the migration decisions of individuals (discussed later, see pp. 93–97).

Another important point made by Lee is that a person's knowledge of conditions in her or his home area is always likely to be more complete, accurate and reliable than that relating to possible places of destination. In the case of the latter, the potential migrant often has to rely on information from secondary sources, such as the media (television, advertising, radio, magazines) or from returned migrants. In both cases the information and image which is conveyed may be incomplete and not altogether accurate. Thus a move to a new location may often be associated with a high level of risk and uncertainty because of deficiencies in the migrant's knowledge about that place, and about the opportunities and conditions which may be found therein.

A further factor which influences the likelihood of people migrating is what Lee termed 'intervening obstacles'. These are potential barriers to migration (such as the cost of travel, the spatial and cultural distance between places, family attachments, personal anxiety, lack of information about opportunities and conditions elsewhere, government restrictions on movement, and so on) which to some people – i.e. those with the means to overcome them: money, contacts, qualifications – may appear slight but which to others may seem insurmountable. We shall return to look at these intervening obstacles later in this chapter.

In spite of its high level of generalization, Lee's model is useful in pointing us in the direction of the factors which should be examined if we are to understand why people move. Thus, we need to be mindful of the circumstances which prevail in the areas from which migrants are moving, and in the areas upon which their movement is focused. We also need to consider the migrant's propensity to move, and the factors which both facilitate and hinder the movement from one place to another. The following discussion seeks to accomplish this with reference to factors associated with the development process, both in source and destination areas, which influence the scale and pattern of migration in the Third World, and particularly that which takes place between the countryside and the city. We start with a brief assessment of some of the factors which may serve to 'push' migrants from their rural homes.

Rural 'push' factors

It would not be difficult to draw up a systematic list of the various forces and conditions in rural areas which underpin people's decisions to move to new locations. Factors such as population growth, land shortages, low levels of agricultural productivity and income, and a weak non-agricultural sector have all exerted a powerful influence on the incidence of rural emigration. In reality, however, it is very difficult to isolate the influence of one or the other factor: these and other influences, including those which emanate from the city, are interrelated and exert

Plate 4.1 The drudgery of rural life. Many migrants in the Third World are only too keen to escape the hardships and uncertainties which are often associated with life in the countryside.

a simultaneous influence on the migration decisions of individuals and communities.

One of the most commonly used explanations for out-migration from rural areas is the high rate of rural population growth in many Third World regions. Whilst rural populations are in general growing at a slower rate than in urban areas because levels of urbanization are still quite low, particularly in Africa and Asia, the numerical increase of population in rural areas is often substantial. The simultaneous occurrence of rural population growth and rural out-migration leads to the convenient conclusion that one is causing the other. This assumption underpinned a number of analyses of migration in the 1950s and 1960s where the movement of population was thought to reflect a surplus of labour in rural areas which was largely a consequence of demographic pressures. Whilst there may be a close link between high population densities and the incidence of migration, as is the case with migration to the Sierra Leonean capital Freetown, population growth is not in itself the main cause of emigration from rural areas. Its effects have to be seen in conjunction with the failure of other processes to cater adequately for the needs of the growing rural populace.

One of the principal needs is for land, and it is pressure on land resources which provides a strong imperative for migration in a great many Third World countries. In parts of Kenya and Uganda, out-migration has been particularly pronounced from areas where available land is no longer sufficient to maintain people's levels of livelihood. In rural Mexico, population growth and the sub-division of plots of land over several generations have resulted in families farming insufficient land for their needs, leading to a growing incidence of migration to Mexico City.

In Latin America the problem is also compounded by the massive concentration of land resources in the hands of a few. This has particularly affected the land-hungry and landless, who have been forced into migration in order to make ends meet. In an analysis of data from sixteen Latin American countries (see Table 4.1), Shaw found that rates of rural emigration were highest in countries where more than half of all agricultural land was held as *latifundia* (more than 500 hectares) and also where more than half of all farms were *minifundia* (less than 5 hectares).

It is not only the amount of land that a family has to farm but also the quality of that land which may have an influence on out-migration. The colonization of Amazonia, with its potentially abundant land

Table 4.1 Average rates of rural emigration for sixteen Latin American countries classified according to the distribution of land resources

	Countries with 0–50% land on farms exceeding 500 hectares		*Countries with 50–100% land on farms exceeding 500 hectares*	
	0–50% farms less than 5 hectares	*50–100% farms less than 5 hectares*	*0–50% farms less than 5 hectares*	*50–100% farms less than 5 hectares*
Rate of rural emigration	0.56	1.48	1.60	2.33
Number of countries	5	5	3	3

Source: Adapted from R. P. Shaw, (1974) 'Land tenure and the rural exodus in Latin America', *Economic Development and Cultural Change*, 23 (1), p. 131.

resources, has in many places failed to provide a permanent solution to land-hunger, and thus migration, on account of the region's unsuitability for many of the kinds of agriculture for which it is being used. In parts of Africa, especially the Horn of Africa and areas fringing the Sahara, where prevailing environmental conditions will only support very low population densities, even very slight demographic or environmental changes have a significant impact upon the rate of out-migration. Furthermore, in many parts of Asia population growth and shortages of cultivable land are combining to push people into marginal ecological zones (uplands, lowland swamps, short-lived riverine islands) in search of land resources, or into increasing the intensity of land use. In both cases, declining soil fertility or occasional environmental disasters may result, adding further to the impetus for people to leave their home areas in search of additional or alternative sources of livelihood.

The characteristics of migration may also be influenced by the availability of land in the home community. The high incidence of landlessness in Latin American countries has led to a much higher incidence of permanent migration to the cities than is the case in many African and Asian countries, where circular migration may be more prevalent because migrants are often able to retain a small plot of land, and through this a social stake in their home communities. This is particularly important in tropical African societies where land is communally held, and where permanent migration may lead to people losing their rights of access to territory and property.

Where people move because of the inadequacy of land resources, rather than because of a total lack of them, it is common for one or two household members to migrate to town to obtain an income with which to supplement the livelihood of those who remain in the village to work on the land. In Africa and Asia there is frequently a division of labour along gender lines, with the women staying behind to tend the fields while the menfolk migrate periodically to town. The reverse is the case in Latin America and in some of the more rapidly-industrializing countries of Asia, where the nature of urban employment opportunities often results in migration streams consisting predominantly of women.

Even where population growth and land scarcity are prevalent, however, it does not necessarily follow that migration will result. Alan Simmons uses the example of two tribal groups in Senegal, the Wolof and the Serer. Even though the Serer experienced much higher population densities and much lower living standards, levels of migration were appreciably lower than among the Wolof. This paradox was explained by the Serer's much stronger religious attachment to their land, which restricted their propensity to migrate, and also to their greater ability to come to terms with the effects of population pressure, by innovating, than was the case with the Wolof. Whilst population growth may thus act as a catalyst for out-migration, it should not be seen as a cause of migration in its own right.

The effects of rural population growth on levels of out-migration have also been compounded by the generally slow and uneven pace of economic and technological change in the rural sector generally and the agricultural sector in particular. The penetration of the market economy into even the most remote corners of the Third World has increased the importance of cash as the main medium for economic transactions. At the same time, however, the capacity of agriculture to satisfy the rising need for cash is constrained by the widespread persistence of inefficient farming practices and a scarcity of investment capital, particularly amongst those groups of farmers who are most in need of support. The limited development of the non-agricultural economy also means that there are few employment opportunities which might enable people to earn a cash income from working locally.

In Indonesia the lack of work is responsible for the migration of more than one million rural labourers to work as seasonal migrants in Jakarta each year after the harvest has been completed and the fields prepared for the new crop. Similarly, in the Kenyan district of Kakamega as many as one-third of the economically active population finds temporary and permanent work in Nairobi and Mombasa, and on large commercial farms in other districts, because of inadequate employment opportunities in the

Plate 4.2 Two-wheeled 'iron buffalo' *(khwaay lek)*. The potential for mechanization to displace large numbers of people from the land in many Third World countries is enormous

district itself. In the face of such circumstances, labour migration thus provides an invaluable source of cash income, especially for poor and land-hungry households which struggle to satisfy even their basic subsistence needs from the land.

If the underdeveloped state of rural areas provides the impetus behind out-migration, particularly from economically depressed Third World regions and amongst relatively underprivileged rural households, then it should follow that attempts to improve the economic performance of the rural sector will have the effect of reducing rates of rural emigration. Whilst the intensification of agriculture and the introduction of modern farming practices has helped to absorb population increases in several areas, notably Java, the Indian Punjab and parts of China, the modernization of agriculture has very often had the opposite effect. For instance, the transformation of a major *hacienda* into one of the most efficient in Ecuador by introducing machinery and improved cultivation techniques had the effect of displacing half of the resident population from the plantation. Elsewhere in Latin America programmes of agrarian reform have released peasant farmers from traditional systems of access to land (particularly 'feudal' systems of tied labour) which had hitherto restricted the scope for migration. Without parallel efforts to raise agricultural productivity and to generate employment opportunities in the rural sector, agrarian reform programmes have added considerable momentum to migration streams between rural and urban areas, much of it permanent because of high levels of landlessness in the areas where they have been implemented.

The need for greater efficiency in capitalist farming has also often resulted in the displacement of large numbers of agricultural workers as mechanical and technology-intensive farming systems have replaced labour-intensive practices. Many thousands of paddy farmers have been displaced from the land in Malaysia as the government has sought to improve productivity in the rice-growing sector by investing in major irrigation schemes and by consolidating fragmented paddy farms. With little work to occupy them locally, landless labourers and former tenants have been forced into migration to urban areas or into starting afresh on one of the country's land development schemes.

The characteristics of rural development in Third World countries, described in Chris Dixon's book in this series, are such that people are thus being displaced from the countryside because in some areas change is too slow to adequately accommodate the growing size and needs of the rural populace, or because in other areas change is too rapid to enable redundant rural workers to be absorbed into other forms of production *in situ*. Migration helps to reduce the number of mouths to be fed, and at the same time may yield an invaluable supplementary source of income for poor, landless or indebted households. Emigration may be a 'last resort' for most rural households, but it is not difficult to imagine the kinds of pressures and problems which would result if out-migration was not an option open to a large segment of the rural populace.

Migration may thus be seen as a kind of 'pressure valve' through which may escape those who might otherwise struggle to survive in an increasingly beleaguered rural sector. We should however be careful not to see migration as the only choice open to rural folk who find themselves faced with such difficulties (nor, for that matter, should we perceive of all rural areas in Third World countries as being beset by the kinds of problems which are outlined above). People may respond by adapting to their changing circumstances, or they may organize themselves politically and economically in order to confront whatever forces are responsible for their plight.

Urban 'pull' factors

Economic motivations underpin the great majority of migratory movements in the Third World (see Table 4.2). We have seen that people may feel compelled to leave their rural home areas on account of limited opportunities for producing goods or earning a cash income to sustain themselves and their families. Given the nature of the development process in many Third World countries, their best chances of obtaining a satisfactory level of livelihood are most likely to be found in an urban centre which may be expected to offer a range of

Table 4.2a Reasons for migration from village communities in rural Peru

Reason	*Percentage of respondents citing reason*
To earn more money	39
To join kin already working	25
No work in the villages	12
Work opportunities presented themselves	11
Dislike of village life	11
To be near the village and family	11
To support nuclear and/or extended family	9
Poor	8
To pay for education	7

Source: Julian Laite (1988) 'The migrant response in central Peru', in Josef Gugler (ed.), *The Urbanization of the Third World*, Oxford: Oxford University Press.

Table 4.2b: Principal reasons for migration from village communities in north-east Thailand

Principal reason	*Number of respondents citing reason*	*Percentage of respondents citing reason*
To earn more money for the household	138	52.9
To earn more money for self	57	21.8
To earn more money for parents	31	11.9
To further education	12	4.6
To earn money to build a house	10	3.8
To earn money to invest in farming	4	1.5
For fun	3	1.1
To earn money to purchase land/ land title	2	0.8
To earn money to repay a debt	1	0.4
To earn money to pay for hired labour	1	0.4
To see Bangkok	1	0.4
To earn money to get married	1	0.4
Total:	261	100.0

Source The author.

employment opportunities in manufacturing, construction, commerce and the service industry, together with a diverse range of social amenities and attractions. An awareness of these opportunities may in itself provide a powerful motivation for migration. It does not necessarily follow, however, that the people who are drawn towards the city by the lure of employment and other opportunities will be the same people who feel compelled to leave their home communities because of the lack of economic opportunities there.

During the 1960s and 1970s, the principal cause of rural–urban migration was argued to be the much higher wages and more varied employment opportunities which were available in the city. Patterns of migration coincided very closely with the nature of such wage and employment differentials, and were also observed to change as these differentials changed. In Tanzania, rates of migration generally increased as a result of improving urban income levels. In Egypt, migration rates were found to have fallen very sharply as a result of improvements in income levels in the rural areas from which the majority of migrants originated. In north-east Brazil, wage rates were sometimes higher in the modern commercial agricultural sector than in many urban centres in the region, and these were also partly reflected in a prevailing pattern of rural–rural migration.

Income differentials are quite easy to quantify (see Tables 4.3 and 4.4). However, they do not take into account differences between rural and urban areas in the cost of living, which is often substantial because food, energy, transportation and housing are usually very much cheaper (if not free) in the countryside than in the city. Also, non-cash income (e.g. crops produced and consumed, goods bartered and exchanged) often makes up a large proportion of rural incomes. Migrants must also spend money in getting to the city and in supporting themselves while they search for work. When these factors are taken into consideration the differentials recorded in Tables 4.3 and 4.4 do not look quite so wide. Wage rates may indeed be higher, but a migrant's ability to save money while working in town may be much less than many migrants expect when they first move to the city. Furthermore, average wage rates may be meaningless as a basis for comparison between rural and urban areas: when wage rates for migrants' particular skills, qualifications and experience are compared, the differential may be very narrow indeed.

Also during the 1960s and early 1970s, it was widely held that the higher the level of urban unemployment, the lower would be the

Table 4.3 Average salaries in agriculture and manufacturing in selected Latin American countries, 1971

Country	*Type of salary*	*Monetary unit*	*Agriculture*	*Manufacturing*	*Agricultural as a percentage of manufacturing wages*
Costa Rica	Hourly	Colón	1.77	3.54	50
Panama	Weekly	Balboa	12.50	29.90	42
Paraguay	Monthly	Guarani	5814.00	7604.00	76

Source: Alan Simmons, Sergio Diaz-Briquets and Aprodicio A. Laquian (1977) *Social Change and Internal Migration*, Ottawa: International Development Research Centre.

Table 4.4 Urban/rural and non-agricultural/agricultural ratios in per capita incomes, selected countries (1960s and 1970s)

	Urban/rural per capita income ratios		*Non-agricultural/ agricultural earnings*
Bangladesh	2.70:1	Ghana	2.28:1
Sri Lanka	2.07:1	Sri Lanka	2.70:1
India	1.67:1	India	2.89:1
Philippines	1.92:1	Malawi	4.02:1
Thailand	2.05:1	Tanzania	2.23:1
Zambia	9.38:1	Zambia	2.93:1
Uganda	3.24:1	Kenya	2.63:1(men)
Brazil	2.73:1		3.73:1(women)

Source: Adapted from M. Lipton, (1977) *Why Poor People Stay Poor: A Study of Urban Bias in World Development*, London: Temple Smith.

likelihood of cityward migration. In reality, a quite significant proportion of rural migrants spend their first few weeks, or even months, in the city without work. Furthermore, in spite of the supposed influence of higher wage rates, many recently arrived rural migrants join the floating mass of urban poor. The parallel existence of large-scale urban-bound migration and rising levels of urban poverty and unemployment led analysts to reappraise the reasons why people were migrating to the cities from the countryside. Either their migration was highly irrational, or migrants were poorly informed about the harsh realities of life in the city. More plausibly, it may be that migrants who are faced with poverty and unemployment are not those people who were attracted to the city by its better prospects but those who had been forced to leave their home areas because of landlessness, poverty and exploitation.

Michael Todaro was one of the first to suggest that the paradox of migration in pursuit of higher wages and better employment prospects on the one hand and urban deprivation on the other could be explained by taking a longer-term view of why people move towards the cities. People were moving not so much because of the *immediate* prospect of improving their living standards but because of the greater likelihood of *eventually* obtaining a good job and an acceptable level of income. Thus people were willing to endure short-term difficulties in the hope of better prospects for economic gain and improved welfare in the longer-term, even if only for their children and not themselves. Expected wages were discounted against the prospects of remaining unemployed for any length of time.

In much of tropical Africa during the post-independence period high levels of urban unemployment amongst the migrant population were explained by people's willingness to wait for scarce but well-paid urban jobs which, because of the protection of government legislation, offered very good income levels and working conditions. In Sulawesi, Indonesia, research has shown that a significant proportion of migrants who remain unemployed for extended periods come from more wealthy rural households and are thus better able to endure unemployment while they wait for a job which is commensurate with their background and qualifications.

It is generally the case that migrants' prospects of economic advancement *are* indeed better in the city than in the village. In Argentina more than three-quarters of a sample of migrants to Buenos Aires considered themselves to be better-off economically than before their move to the city. Similarly, even some of the poorest street traders in Jakarta, who could hardly afford to feed their families, reported that they were better off in the city than they had been back in the village: income levels for many had increased by as much as two-thirds. In New Delhi it was found that it was not wage differentials which accounted for better incomes amongst poorer migrants from rural areas but the fact that they could find perhaps twice as many days work in the city, resulting in income levels which were 2.5 times those enjoyed in the village. If we add to this the fact that they can generally expect to enjoy better welfare facilities in town (education, health care, housing, public amenities), then their move to the city appears to be entirely rational. Migrants' knowledge and awareness that conditions and prospects are generally better in urban than rural areas helps to explain why such large numbers of rural migrants are attracted to the major urban centres of the Third World. Given the choice, many would prefer to be unemployed in the city, with its better longer-term prospects, than underemployed in their

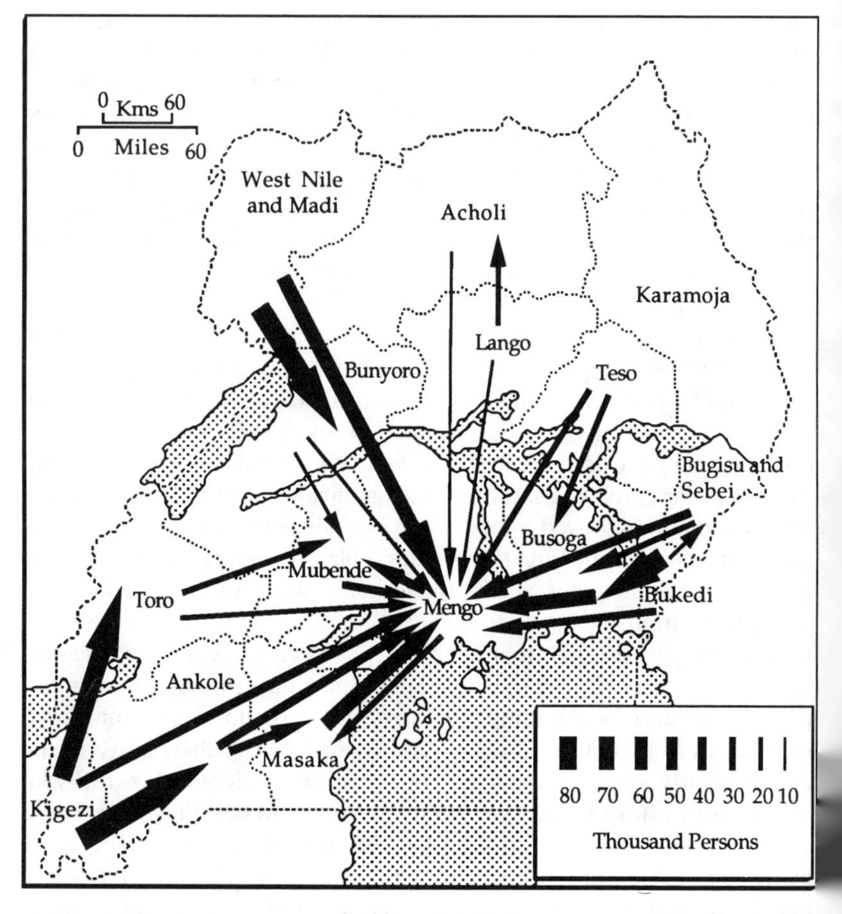

Figure 4.2 Uganda: inter-regional migration flows greater than 10,000 people
Source: I. Masser and W. T. S. Gould (1975) *Inter-Regional Migration in Tropical Africa*, London: Institute of British Geographers, Special Publication no. 8, p. 52.

home village. Conversely, others would opt to be unemployed in the village because there they can rely on the support of family and friends. This further explains why some people migrate, however irrational their movement may appear in the short-term, while others prefer to stay put.

The discussion thus far has focused on differences in income and employment opportunities between rural and urban areas. Similar differentials can be identified on a wider scale between dynamic and economically depressed regions. Figures 4.2 and 4.3 show clearly that in addition to the predominant focusing of migration streams on the capital cities of Uganda and Sri Lanka, there is also a pronounced tendency for the streams to converge on these countries' more prosperous and dynamic core regions and away from the economically backward peripheral regions. This reflects the strong gravitational pull of the capital region. In Indonesia too the greatest level of mobility is found in and towards the economic 'boom' provinces not only in the centre (Jakarta and North Sumatra) but also in dynamic peripheral regions (South Sumatra, Riau and East Kalimantan, where development has occurred in association with the exploitation of natural resources). Conversely, the highest rates of out-migration occur from the economically stagnant provinces. Similar patterns of inter-regional movement are also found in many Latin American countries.

Wage and employment differentials must also be viewed in relation to other factors which influence the migration process. In Ghana it was found that rates of rural–urban migration continued to increase even though the income gap between the countryside and the city began to close rapidly during the mid-1980s as a result of government programmes of support for the rural sector (particularly through increasing prices paid to cocoa producers). The most likely explanation for this paradox is that the government's policies were not affecting all farmers equally but were being particularly beneficial to the larger commercialized farms. Also, it should not be forgotten that people do not migrate only because of perceived economic differentials but also for a wide variety of non-economic reasons.

The so-called attraction of the 'bright lights' of the city is one such non-economic factor which has been widely used to explain the incidence of rural–urban migration in Third World countries. Potential migrants are made constantly aware of the attractions of the large city through such media as advertising and soap operas on television, and through the stories which are relayed by migrants who return to their home communities – traditionally, complete with imaginative interpretations and embellishments. The city is thus portrayed and seen as *the* place to find fun and excitement, and in the mind's eye of young impressionable villagers contrasts sharply with the generally slow and unexciting pace of life in the countryside. In the Ivory Coast it was found that the positive image of the city which was conveyed by returning migrants acted as a strong encouragement to others to move in the same direction.

In this sense, the non-economic attractions of the city may not so

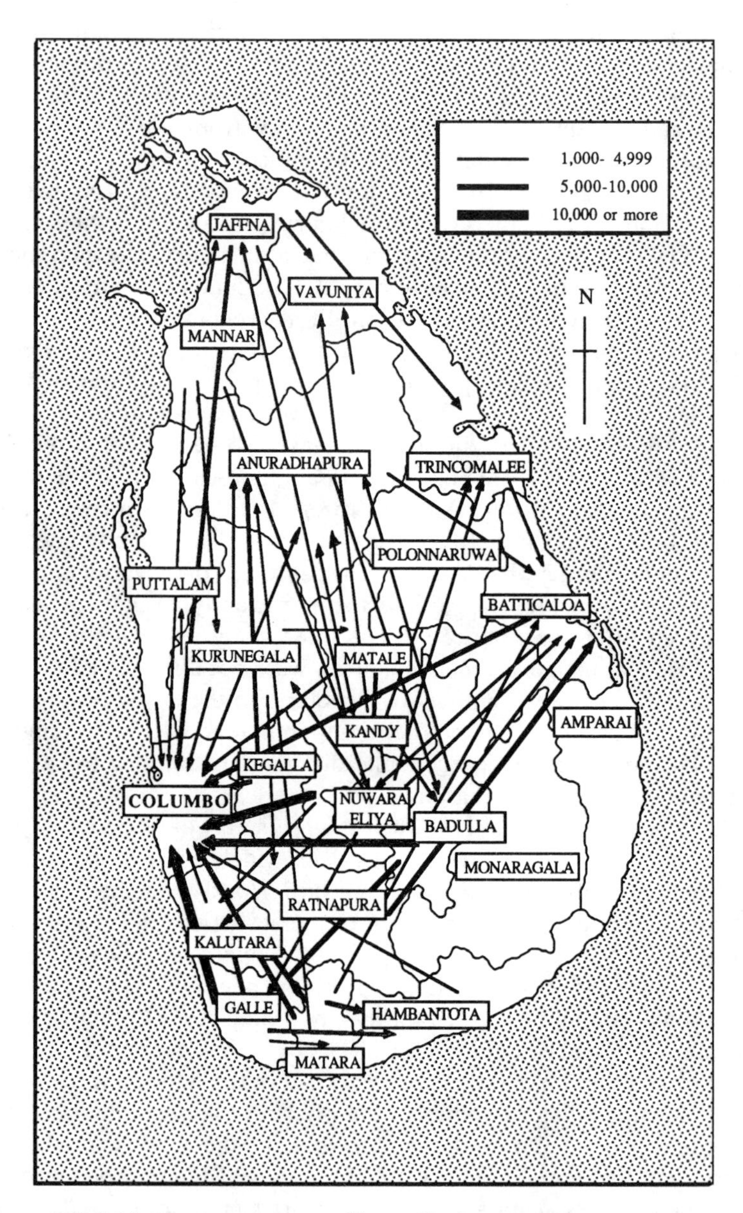

Figure 4.3 Sri Lanka: net streams of inter-district migration
Source: A. S. Oberai (ed.) (1983) *State Policies and Internal Migration: Studies in Market and Planned Economies*, London: Croom Helm, p. 90.

much determine the *incidence* of migration (although migrants do sometimes leave their home communities for the sole purpose of seeking fun in the city), but may have a very strong influence on the *direction* in which the move takes place. If a migrant has to choose between several alternative destinations, all of which offer much the same range of economic opportunities and prospects, the chances are that he or she will opt for the location which offers the preferred social environment. In this way it is easy to understand why such a large proportion of rural–urban movements tend to focus on the capital city, where such social opportunities tend to be so much more prevalent.

It should none the less be borne in mind that many of those who are newly arrived in the city, other than those from the middle and upper echelons of society, do not have the financial means to avail themselves of the many attractions that the city has to offer. For many, also, the hustle and bustle and danger of the city provide a very intimidating and frightening environment which holds far fewer attractions than the home community.

It is therefore clear that there are a wide range of factors in both rural and urban areas which influence the incidence and direction of population movements in the Third World. Whilst such meso-level factors yield a clearer understanding of the precise reasons why people move than is the case with the macro-level perspective, they none the less also assume a general homogeneity of migration streams and a basic inevitability of movement. As we have seen in earlier chapters, streams of movement between rural and urban areas are often highly differentiated, and consist of people who have been forced by circumstances into leaving their rural homes and others who have opportunistically moved to take advantage of the prospects the city potentially offers. The characteristics of these two types of migrants, and the likely effects of their movement, may be quite fundamentally different. We have also seen that, however pervasive the meso-level influences might be, the majority of people decide not to move. It is only by examining the micro-level influences on the migration decision-making process, i.e. those operating at the level of the household or the individual migrant, that we can obtain a more or less complete understanding of why some people move and others do not. It is important to note that not everyone needs to migrate, not everyone wants to migrate and, as we shall see below not everyone is able to migrate.

Case study D

Filipino migrant workers around the world: the labour trade

Since the 1960s there has been a dramatic increase in the exodus of Filipinos to work abroad, particularly for short-term contract workers. Between 1975 and 1985 well over one and a half million Filipinos travelled to find work abroad. The increase occurred during one of the most dramatic periods of political and economic crisis in the country's turbulent history. About 70 per cent of Philippine families now live below the government poverty line. Stiff repayment conditions imposed on the government by foreign lenders for a foreign debt of over US$25 billion, make it unlikely that living standards will improve before the end of the decade. Widespread disenchantment with the regime of President Ferdinand E. Marcos, who for twenty years presided over a catastrophic decline in the country's economic performance, led to his fall in February 1986. The new government, led by President Corazon Aquino, faces enormous problems in re-establishing the economy and public confidence.

Under President Marcos, the government actively promoted the trade in labour. It sought to take control over the profitable business of recruitment and passed laws which required Filipinos working abroad to remit as much as 80 per cent of their earnings through Philippine banks, and which taxed them and the banks in the process. Between 1977 and 1983 Filipino migrants sent home more than US$3.5 billion – about US$70 for every man, woman and child in the country. Migrant workers contribute more foreign exchange to the Philippine economy than traditional exports of sugar, minerals and wood products.

The great majority of those who go abroad are in the prime of their working lives. More than 80 per cent are aged between 25 and 44. They are also educated: over 80 per cent have completed high school, many have college or professional qualifications, and the great majority have at least one year's work experience. They are therefore among the country's most active and skilled people – a fact which raises fundamental questions about the real benefits of a trade which now affects the lives of almost every Filipino family.

Case study D *(continued)*

The material well-being of many families has undoubtedly been improved because one or more family members have gone abroad to work. The country's economic crisis makes it desperately hard for the majority of skilled and semi-skilled workers to support their children. For this reason, emigration offers many the only hope of breaking out of the poverty trap.

The export of Filipino labour does not make a desirable or positive impact on the Philippine economy in general. The idea that it brings in much needed foreign exchange to offset the monetary crisis is only partially correct. It is also false that labour export has solved the problems of unemployment. The labour export programme harms the domestic economy by syphoning off skilled workers who cannot be replaced and who have a crucial contribution to make to the country's economic development. At present, the Philippines is losing its most precious skills, and shortages are already evident in important sectors of the economy. To make matters worse, many of those working abroad are not even acquiring new skills, but are becoming 'de-skilled'. However many dollars and cents are earned from the labour trade, the long term costs incurred by the country are likely to outweigh any benefits.

Source: Catholic Institute for International Relations (1987) *The Labour Trade: Filipino Migrant Workers Around the World*, London: © CIIR, pp. 6, 128.

Micro-level factors

The establishment of networks of contact with urban areas over a period of several generations may be of central importance in both initiating and facilitating migration from rural areas. The city contrasts markedly with the village in terms of size, environment, pace of life, modes of economic and social activity, and so on. Where migration takes place over large distances, there may also be a sharp cultural and linguistic divide between source and destination areas. First-time migrants may be very apprehensive prior to moving to the city, and they may feel quite alienated and intimidated during their initial period there. It may take months or even years before the migrant becomes sufficiently street-wise to be able to cope with the demands and pressures of urban life.

In such circumstances it is quite rare for people to move to a destination about which they know little and in which they know virtually no-one. Most will move to a location where they have connections, usually in the form of people from their villages of origin. A number of studies have described the important role that these urban contacts play in finding work for migrants from their home communities, and in providing shelter, finance and social support for the newly arrived migrants. A sample survey of rural migrants in Bombay revealed that more than three-quarters already had one or more relatives living in the city, from whom 90 per cent had received some form of assistance upon arrival. Similarly, a survey of rural migrants to Lima from the Peruvian Highlands has shown that some nine-tenths of migrants move knowing that there will be someone in the city who can accommodate them for the first few days and nights, and around half have a job waiting for them when they arrive, courtesy of their contacts in the city.

A study of rural–urban migration in Ghana found that, because of the network of contacts which existed between urban and rural areas, less than one-fifth of migrants had to arrange their own shelter or had to search for accommodation alone. In Papua New Guinea, on the other hand, the majority of migrants who end up squatting on public land in Port Moresby have no established ties with, and contacts in, the city.

In this way it is easy to understand the phenomenon of 'chain migration', where subsequent waves of migration follow in the same direction as those of earlier, pioneering migrants from the same communities. In north-east Thailand it is common to come across villages which specialize in particular forms of migration, and where the majority of migrants from individual villages work in the same occupations and live in the same parts of Bangkok. This reflects the importance of the information which is sent back to the village by migrants in town, and the supporting role that urban-based migrants play in encouraging and facilitating the movement of new migrants to town. In this sense, rural households which cannot easily plug into such networks of information and support are much less likely to send migrants to work in town than those that can.

A criticism of the macro- and meso-level interpretations of migration is that they see movement as a passive response to a variety of stimuli. They also tend to view rural source areas as an undifferentiated entity.

When we view migration decision-making at the level of the individual or household, however, this becomes a gross oversimplification of the true situation. In reality, no two people in the countryside are faced with an identical set of circumstances and thus people cannot be expected to respond to various structural and developmental stimuli in the same way. Differentiation between rural households takes a variety of forms, including variations in levels of income, size of land-holding, the size of the household, stage in the life cycle, levels of education, cohesiveness of the family unit, contacts in and knowledge about other locations, and so on. These and many other variables may each have a significant influence not only upon whether or not people decide to leave their home areas but also upon the characteristics of their migration should they decide to move.

A household's level of disposable income, which may itself be a function of the amount of land owned and levels of agricultural production, and which in turn may influence such factors as education and qualifications, will have a significant bearing upon its ability to afford the transportation and other support costs of the migration which Lee has identified as one of the more important 'intervening obstacles' which condition a person's ability to translate 'push' and 'pull' influences into actual movement. The costs of migration may also vary significantly depending upon whether the migrant is willing to accept the first job and form of shelter that becomes available, or whether he or she prefers to wait until suitable employment and accommodation are found. Where the principal aim of the migration is to improve one's level of education or training, the cost of supporting the migrant in town may be considerable. Clearly not all rural households are able to afford the costs of migration to the same extent, and thus not all are in a position to exploit the economic and social potential of migration to the full. By continuation, shortages of disposable income may also restrict to a much greater degree the migrant's range of choice in terms of when to move, where to go and what forms of employment to accept than will be the case with migrants who can afford to be more selective and choosy.

A range of non-economic factors also have an influence on the propensity or ability of individuals to migrate, and upon the characteristics of their migration. Family ties and commitments may determine whether or not someone is able to migrate, and may also influence who from a family unit is most likely to take on the responsibility of migration. The stage in the life cycle is thus crucial. There are few restrictions on the mobility of unmarried migrants (although in some countries social constraints and customs may make it more difficult for young women to migrate than young men), which explains why the vast majority of migrants in Third World countries are aged between 15 and 25 years. A married couple with young children may face more constraints on migration, unless the entire family moves or the family is willing periodically to be split by the migration of one or other of the parents (typically the father). There may then be fewer constraints in the home area on the migration of older people, but their movement may be restricted by the limited range of employment opportunities open to them in urban centres, and by their own disinclination to migrate.

A remarkable pattern of 'relay migration' has been identified in a study of the Mexican village community of Toxi where, at different stages in a family's life cycle, different people take responsibility for migration. When a family's children are very small, it is usually the father who engages in seasonal forms of migration which keep him away from the village for only a few months at a time. As the children get older, so they take over responsibility for migration, starting with the oldest children. Responsibility then passes systematically to younger siblings as the older ones marry and raise their own families. Throughout the course of a generation, relay migration is thus responsible for maintaining continuity of income-generation from outside the community whilst at the same time minimizing disruption to the family and to the household economy.

The end-product of these various forms of social and economic differentiation in rural areas are similarly highly differentiated streams of migration to urban centres. Some go, others stay. Those that leave are typically younger, better-educated and, arguably, more innovative and dynamic than those who stay behind, and thus the selectivity of the migration process may be resulting in a form of 'brain drain' or selective depopulation of rural areas (see Chapter 5). Some are able to take full advantage of the opportunities the city has to offer, whilst others are thankful to take whatever opportunities may come along, surviving on the very margins of an urban existence. Some decide that their interests will be best served by moving permanently to a new location, whereas others attempt to make the best of both worlds by migrating periodically to the city whilst at the same time continuing to maintain a stake in their home community.

Whilst a wide cross-section of people from rural areas may, at some stage in their lives, become involved in the migration process, the reasons for their movement, and its characteristics and effects, may vary quite significantly. In such circumstances, it is therefore misleading to suggest that migration represents a straightforward response to the relative influence of rural push and urban pull factors. For some households the factors (psychological as well as economic) which attach them to their home communities may be strong enough to override the draw of the city. For others the negative aspects of life in other places may cause them to rationalize the potential benefits to be obtained by migrating there.

Migration decisions may therefore be seen as investment decisions which are based on a given individual's or household's expected costs and returns from migration over time. Such costs and returns are seen in non-monetary as well as monetary terms. The former include the psychological costs of moving away from the home area and of adapting to a different social and cultural environment in another location. These non-monetary costs and benefits are much more subjectively interpreted by people than economic ones, which are much more easily quantifiable: what may appear as a positive factor to one person may have strong negative connotations to another. This also helps to explain why some people move and others do not, even when both appear to be faced with an identical set of circumstances or stimuli. At the end of the day, much rests on the personality, aspirations, attitudes and motivations of individual people. Thus what we have in reality is a complex interplay of a great variety of factors which impinge upon, and influence, the mobility decisions of people in rural (as well as urban) areas. It is therefore inevitable that the outcome should be a similarly complex set of migration forms and types.

Conclusion

Just as we should be careful to avoid viewing the Third World as a homogeneous and constant entity, so too should we be mindful of the great variety of factors which cause people to leave their homes in search of better opportunities elsewhere and, of course, the complex range of movement types to which the interplay of these factors gives rise. This chapter has sought to introduce, at a variety of scales, a broad cross-section of the factors and processes which give rise to migration in Third World countries. It should be clear that there is a close interplay between the pattern and process of development in Third World countries and the incidence and characteristics of migration. In particular, the concentrated pattern of economic activity, and deficiencies in spreading the benefits of economic growth to all areas, sectors and peoples, underpins the overwhelming majority of population movements in Third World countries. As evidenced in the following chapter, migration also has a fundamentally important role to play in terms of influencing the pattern and process of Third World development.

Because of constraints of space, the previous discussion has mainly examined the economic determinants of migration. It should be remembered, however, that there is a great range of social changes and pressures which is also giving rise to a significant volume of population movement in Third World countries. Changes in the position of women in Third World societies, increasing demand and opportunities for tertiary education, and the changing relationship and division of responsibility between children and parents are but three of several fundamental processes which increasingly underpin the mobility of certain groups in society. Even in these cases, however, there is often an economic motive behind the resultant move.

Whilst we have focused in this chapter on the migration decision-making process, it must also be borne in mind that mobility decisions are just part (in some cases a relatively small part) of a set of decisions whereby an individual or household seeks to improve its livelihood and welfare. Thus, cityward migration should not be seen as the only option open to rural households. As with decisions about migration, these decisions and options are also influenced or constrained by the household's present set of endowments: wealth and assets (land, property, savings), education, skills and experience. This also helps to explain why some people move and others do not: some are better able to compensate for deficiencies in their home areas through other means.

Key ideas

1 The unevenness of the development process provides the most powerful explanation for the high incidence of population migration in Third World countries.
2 Whereas some people may move towards a new destination with little knowledge of what awaits them there, the existence of long-established networks of information and contact with the major urban destinations means that the majority of migrants in the Third World are fully aware of the opportunities and difficulties which will face

them in the new location. These networks are also responsible for the development of streams of migration.

3 The decision to move is seldom taken by the migrant alone, and may involve family members, friends or members of the wider community. This is because migration may be seen as an 'investment decision' which will involve costs and benefits not only to the migrant but to others in the home community.
4 Rural–urban migration occurs in response to prevailing and expected conditions in the migrant's home area and in one or a number of potential destination places. Such conditions may be perceived differently by different people, which explains why not everyone responds by moving.
5 Non-economic factors may be important not so much in determining the incidence of migration as in influencing the direction in which the migrant moves.
6 Because of the filtering effect of so-called 'intervening obstacles', population movements tend to be very selective in terms of the age, gender, economic status, ethnicity and other characteristics of the people who are able to move.

5
The effects of migration

It should be clear from the earlier chapters that migration in a highly complex process, and that the reasons for and characteristics of population movements are seldom straightforward and predictable. The same is true of the effects of migration: it simply is not possible to state whether, on balance, the overall impact of migration is positive or negative. So much depends upon who is moving, the circumstances which led to the move, the purpose of the movement, the form the movement takes and, of course, the extent to which the migrant succeeds in achieving his or her objectives. We must also consider the question of scale: whilst individual migrants may derive considerable benefits from their movement, the aggregate impact of several thousands of people moving from a particular area may be very damaging.

There is also, of course, no agreement as to what constitutes 'better' or 'worse' as far as the effects of migration are concerned: indeed, the same outcome (for instance, the return of migrants with fresh, progressive ideas) may be viewed entirely differently by two migrant households depending upon the outlook of the head of household. Furthermore, what may be perceived as a beneficial outcome in the short-term may have deleterious effects in the longer-run: the income sent from town may lead to the farm being neglected, for instance, which may be regretted should a household's access to urban employment be restricted for whatever reason.

We should also bear in mind that it is often very difficult to isolate the effects of migration from those of other aspects of the processes of development and change. Thus, for example, the money and ideas that migrants may bring back to their home communities may give the appearance of stimulating agricultural and other economic improvements, but these have also to be viewed in the light of work being done by various developmental agencies in the home community, the effects of national education programmes and other forms of infrastructural investment, changes in market conditions and so on, all of which may contribute significantly to the developmental environment within which the trappings of migration are being used.

The aim of this chapter is to outline some of the ways in which the process of development is being affected by migration and circulation between rural and urban areas which, as we have seen, are often the prevalent forms of population movement in Third World countries. Earlier chapters have suggested that people are, in effect, 'voting with their feet' against the relative stagnation and backwardness which characterizes rural areas in many Third World countries, and are, in the process, seeking to avail themselves of the opportunities and rewards which they perceive to be available elsewhere. If this is the case (which, as we have seen, is far from certain), the central issue to be addressed in this chapter is just how effective migration is in terms of reducing the social, economic and spatial inequalities which often give rise to migration in the first place. In other words who, and which areas and sectors, benefit the most from migration? The discussion will focus on the effects of rural–urban migration on the areas from which people move, the major destination places and also upon the migrants themselves.

Effects on source areas

Figure 5.1 summarizes some of the main costs and benefits of migration and circulation for the rural areas from which the majority of migrants originate. The diagram highlights the main factors which determine how rural areas are affected by migration – namely, the two-way transfers of labour, money, skills and attitudes. Much of the discussion of these factors which follows is based on a number of individual village studies which provide the main source of information on the effects of migration on rural areas.

Labour

Whatever the reasons for migration, it inevitably results in the periodic or permanent absence of people from their home areas. Unless the

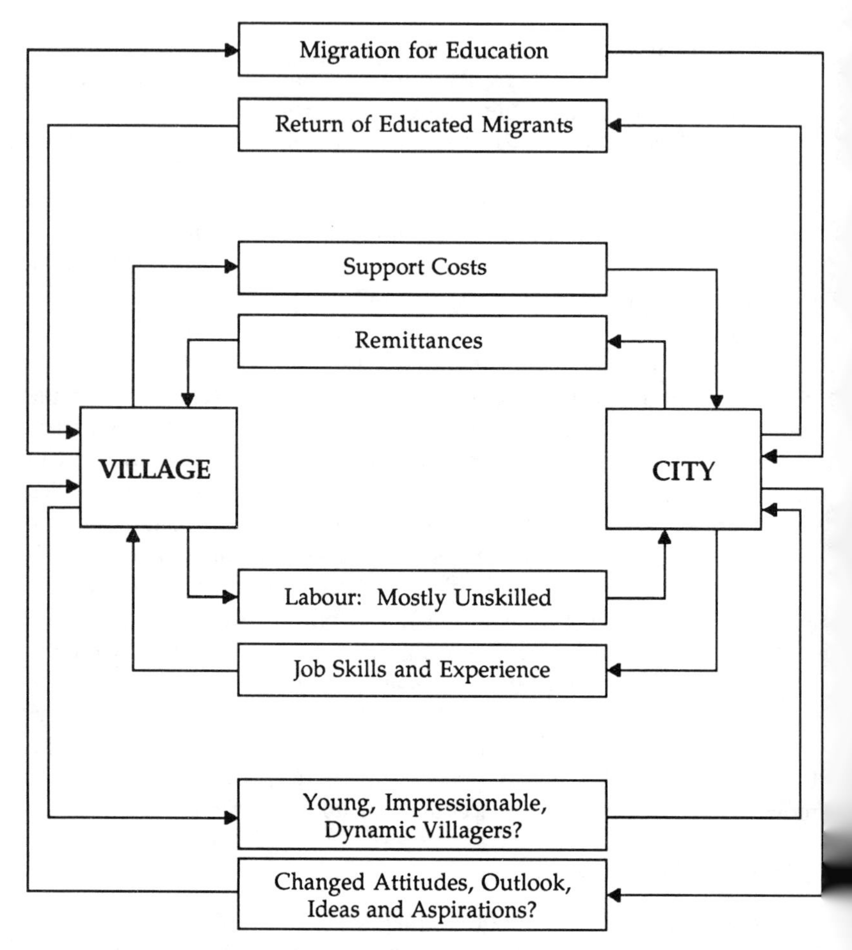

Figure 5.1 Simple model of the costs and returns from migration

absentee is not economically active (e.g. for reasons of age, disability or lack of opportunity), it follows that migration also draws potentially productive labour away from the source community. Migration may therefore lead to a reduction in a household's ability to make the fullest use of productive resources such as land. However, the effects of migration in this regard are seldom so straightforward. There are a number of factors which determine the extent to which migrant households are affected by the periodic movement of labour to an urban centre. Some of the more important of these will be outlined below.

The effects of out-migration on farming will depend, first, upon the volume of movement in relation to the availability of land and the size of the population in the home community. Thus if labour is scarce, the absence of one or more household members may have a significant negative impact upon the household's ability to maintain levels of production and income from agriculture. If labour is relatively plentiful in relation to the amount of land to be cultivated, migration may have the effect of raising labour productivity amongst those who remain.

During the 1950s and 1960s, it was a widely held view that migration was helping to relieve population pressure in rural areas by moving surplus labour from the countryside, where it was underemployed, towards the cities where it would be more productively and efficiently employed in industry and commerce. Rapid population growth in rural areas, set against diminishing land resources and rising person/land ratios, was considered responsible for the growing misery of the landless and the land hungry. Migration thus represented an important means by which person/land ratios were being stabilized, or even reduced.

We saw in Chapter 4 that population pressure and land scarcity are important factors in causing people to move to urban areas, particularly in the countries of South and South-East Asia where there is little cultivable land which has not already been cleared for agriculture. However, only a relatively small proportion of the total volume of movement in the Third World takes place for this reason. Where it does occur, it is doubtful whether agricultural production is suffering. Village studies in Papua New Guinea suggest that person/land ratios are such that up to one-third of male villagers would have to leave before agricultural production would be adversely affected. Studies in Central Africa similarly claim that subsistence production would remain unaffected unless half of the male population was withdrawn through migration.

In parts of Sierra Leone, out-migration was helping to ease pressure on land resources and was actually raising levels of labour-productivity as people strived to maintain existing levels of food and cash crop production. A similar outcome has been observed in the Indian Punjab. Furthermore, the loss of labour through migration may create employment opportunities for others in the source community, and may also push up local wage rates. In cases of extreme poverty, migration may also be welcomed for easing pressure on a household's scarce food resources – leaving one or several fewer mouths to be fed.

In contrast, studies in the north-west Malaysian state of Kedah and the Dominican Republic have shown that migration has resulted in quite severe shortages of labour, and that agricultural activities have had to be scaled down, changed to less labour-intensive crops, or even abandoned altogether. (Once again, however, we cannot completely isolate the effects of migration from other influences such as, in the Malaysian case, the low price farmers receive for paddy rice, which may also have influenced the initial decision to migrate.) In the Yemen Arab Republic, the loss of labour through migration was responsible for the neglect of essential maintenance activities, which in turn led to the collapse of terraces and the diminished efficiency of irrigation facilities.

In the Senegalese village of Soninke extensive migration was responsible for land remaining uncultivated, even though workers were brought in from outside the village to tend the fields in the absence of the village menfolk. Finally, the volume of male out-migration in Zambia had the effect of slowing the rate of land clearance and land maintenance, which in turn was responsible for the overcultivation of existing land resources, leading to problems of declining soil fertility and increasing soil erosion.

A second factor which influences the impact of migration on labour-availability in rural areas is the timing and duration of a migrant's absence. Where irrigation facilities are not available to reduce the dependence of farming on rainfall and other climatic influences, agricultural activity in many parts of the Third World may be characterized by distinct peaks and troughs, especially in terms of labour demand. In such circumstances, the timing and duration of population movements may be important in determining the effects of people's absence from their home areas. Where movement takes the form of seasonal circulation, and thus is timed to coincide with the agricultural cycle (see Figure 5.2), it is less likely to have a disruptive effect on production than where migrants are away from the village during peak periods of faming activity.

Such a view is, however, slightly misleading in that it presumes that labour is not productively employed during supposedly 'slack' periods in the farming calendar. In reality, the time between planting and harvesting, and particularly between harvest and land preparation for the following growing season, may be when essential maintenance activities take place. Where such activities are neglected because of migration, problems will almost inevitably occur in the longer-term, as in the case of the Yemen Arab Republic mentioned above. Crucially, it is also during the slack farming periods that rural households may turn

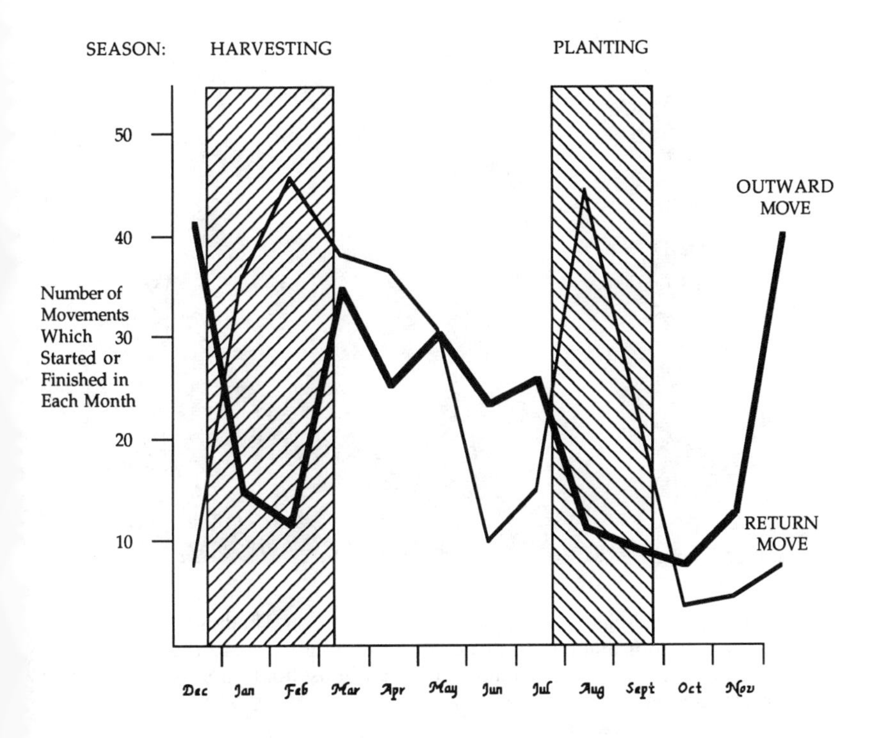

Figure 5.2 The timing of outward and return movements in relation to the agricultural cycle in north-east Thailand, 1968–80
Source: The author.

their attention to non-farm activities, such as cottage industry, which may provide a valuable supplement to agricultural earnings. Thus in a number of Third World countries there are few, if any, times in the year when labour can be considered to be wholly redundant.

A third factor concerns the household's ability to compensate for the loss of labour. We have already seen that households can react 'negatively' to temporary or sustained shortages of labour by leaving land uncultivated or by switching production to less labour-intensive crops (and also by switching from cash crop to food production as a form of subsistence security). More positively, they may hire workers to do the tasks formerly undertaken by the migrant(s), or they may use labour-saving machinery.

Plate 5.1 Labour exchange *(long khaek)* in north-east Thailand. Reciprocal exchanges of labour among kin or neighbourhood groups are rapidly disappearing because migration interferes with households' ability to fulfil their labour obligations.

Levels of agricultural production may be maintained by other processes. Kinship groups amongst the Mambwe society in Zambia have developed labour exchange schemes whereby non-migrating males help out with various tasks which had previously been undertaken by men who now migrate to work in town. Paradoxically, in several South-East Asian societies migration has had precisely the opposite effect, making it increasingly difficult for migrant households to fulfil their reciprocal obligations under traditional schemes of cooperative labour. The resultant switch to hired labour is partly responsible for the rapid decline of this traditional system for overcoming periodic labour shortages.

A fourth factor in determining how migrant households are affected by the loss of labour revolves around the characteristics of the mover. Professional people and those with skills (craftspeople, doctors, teachers, etc.) are much less easily replaced than unskilled labourers. Furthermore, the loss of skilled workers and entrepreneurs may have the effect of reducing employment opportunities for others who remain in the home area. In the Dominican Republic out-migration is highly selective in respect of the skills, productivity and dynamism of those involved, to the extent that the unskilled people who are left behind are unable to adequately compensate for their absence by taking over their roles and tasks.

The tendency for migration to involve a relatively narrow range of economically active people in terms of age, and to some extent also gender, often necessitates drawing people into the rural workforce who might otherwise not be so centrally involved, including the old and very young. A common response to the regular or prolonged absence of male household members (at least, in countries where migration streams are still dominated by men) is for women to take on a greater burden of work and responsibility, both inside the household and in farming or business. The book by Janet Momsen in this series has shown how women in eastern and southern Africa tend the fields and run the households while their husbands are working away from the village. Levels of female participation in the rural labour force are thus much higher than they might otherwise be. In the Indian state of Maharashtra, Muslim wives of male migrants took on the culturally unfamiliar role of controlling household finances, supervising the farm and directing the education of their children.

Thus, in some societies, migration may be responsible for changing the sexual division of labour quite fundamentally. In Pakistan, by contrast, the absence of male heads of households led to other males within the extended family being brought in to run the family farm, ahead of the absentee's wife or other female household members. There is also little evidence that increased female participation in the workforce was leading to innovation and rising levels of efficiency. In the main womenfolk still have to refer to the male head of household before major production decisions can be taken, even when he is absent from the village for months or years on end. Thus rather than leading to a process of social change whereby women are being removed from the restrictions of their traditionally defined roles, their more prominent position in society tends only to be a temporary phenomenon, with women reverting to their traditional roles once the menfolk have returned from their migration.

By contrast, the greater preponderance of female migration in many Latin American countries places a greater burden on male non-migrants to fill the several domestic and economic roles previously undertaken by the migrant women. A further, wonderfully ironic twist to the traditional experience of uneven gender burdens and benefits from migration, and also the infidelity of migrant males, concerns the migration of Thai men to work in the Gulf. Several stories were circulating in north-east Thailand about cases where the wives of migrant workers had had affairs with other men during their husbands' long absence (typically two years). Their husbands had regularly and reliably remitted money back to their wives, who had squandered it on their lovers or had invested it in building a new home and a new life with their new partner. The first time the migrant may have been aware of these goings-on was when he returned to his home village to find neither spouse nor savings.

Cash transfers

Even where households are not able to juggle or substitute labour inputs, the money that is sent back to the village from town would appear to more than compensate for the loss of labour resulting from migration. Whilst there is plenty of evidence that remittances are often substantial in aggregate terms, especially from international migration (see Table 5.1), and that they also make up a significant proportion of the total income of migrant households, it is far from certain that they adequately and directly counterbalance the costs of migration (financial as well as human).

Table 5.1 Overseas remittances relative to merchandise exports and GNP

	Year	*Total remittances (US$ million)*	*Remittances as a % of: Merchandise exports*	*GNP*
Bangladesh	1981	377	53.0	3.4
India	1980	1,600	19.9	1.1
South Korea	1980	1,292	7.4	3.9
Pakistan	1981	1,900	69.9	8.8
Philippines	1980	774	13.5	3.1
Sri Lanka	1980	137	12.7	3.6
Thailand	1981	450	7.2	1.2

Source: Derived from United Nations, Economic and Social Commission for Asia and the Pacific, (1987) *International Labour Migration and Remittances Between the Developing ESCAP Countries and the Middle East: Trends, Issues and Policies*, Bangkok: UNESCAP, Development Paper no. 6.

Given the reasons for migration which have been outlined in earlier chapters, it is not surprising that migrants generally save and remit a significant proportion of their urban earnings. In Kenya, for example, a large sample survey revealed that some 89 per cent of male migrants in Nairobi regularly sent money back to their home communities, and that on average remittances constituted about one-fifth of their city earnings. In Indonesia, almost all temporary migrants sent money to their families back in the village, and for many remittances constituted around 50 per cent of their city earnings. Viewed from the receiving end, studies in Western Samoa have shown that in some villages remittances make up on average 58 per cent of the total cash income of the village. In one such village, Sa'sai, 84 per cent of the population depend on remittances for around three-quarters of their income. Proportions vary significantly, however, rendering somewhat meaningless any attempt to extrapolate these figures beyond the communities to which they refer. A survey of sixteen villages in India revealed that only a little over a quarter of migrants sent remittances, although this relatively small proportion was probably accounted for by the quite high incidence of rural–rural circulation.

The amount of money sent from town tends to be higher amongst the more wealthy households, principally because of their better contacts and education levels, and thus also urban income levels, but remittances generally make up a higher proportion of total income for poorer households. It may also be the case that the uses to which remittances are put will vary according to the economic status of the migrant's household. Wealthier households may be expected to invest the money sent by migrants in various forms of productive and non-productive enterprise, whereas poorer households may be obliged by their circumstances to give greater priority to satisfying their basic consumption needs.

In assessing the impact of remittances on rural source areas, we must bear two points in mind. First, it is very difficult to isolate the effect of remittances from other sources of income. Unless some form of *target migration* is embarked on, where the move to town is used to raise the money to purchase a particular item, remittances will generally be seen to form part of the recipient household's total cash income. This being the case, the impact of remittances can only really be gauged in terms of the effect of raising the household's aggregate income. Second, there is generally a close correlation between the volume, use and impact of remittances on the one hand, and the reasons for and purpose of the migration on the other. Thus in assessing the impact of the financial transfers from town, we need to be mindful of the factors underpinning migration decisions which were outlined in Chapter 4.

Regarding the use of remittances, the majority of village studies

Table 5.2 Expenditure patterns of remittances to Pakistan, 1979

Expenditure category	*Average expenditure per migrant (US$)*	*%*
Consumption	*1,820*	*62.2*
Recurring consumption	1,668	57.0
Marriages	69	2.4
Consumer durables	83	2.8
Real estate	*633*	*21.6*
Construction/house purchase	355	12.1
Home improvement	66	2.3
Commercial real estate	167	5.7
Agricultural land	45	1.6
Investment/savings	*379*	*13.0*
Agricultural investment	97	3.3
Industrial/commercial investment	240	8.2
Financial investment/savings	42	1.4
Residual	*93*	*3.2*
Total	*2,925*	*100.0*

Source: Calculated from Charles W. Stahl and Fred Arnold (1986) 'Overseas workers' remittances in Asian development', *International Migration Review* 20 (4).

conclude that 'consumption' on essential and luxury items greatly outweighs 'investment' in productive forms of enterprise such as agriculture and local businesses (see Table 5.2). Studies in India, Papua New Guinea, Thailand and in parts of East Africa have shown that up to 90 per cent of the income derived from remittances is used for what might be considered 'consumption' forms of expenditure (although it is difficult to fit many forms of expenditure, such as education, into one or the other category). This is not surprising, given the main reasons why migration is taking place. Poverty and a lack of income-earning opportunities in rural areas means that many people are forced to look outside their home areas in order to supplement their livelihood. It is thus entirely logical that remittance income should be used to offset the problems of poverty and to satisfy the consumption needs of migrant households. Indeed, it would be quite illogical for them to seek to satisfy their needs *indirectly* by first investing remittances in farming, unless the returns on this investment were substantial and could be guaranteed which, of course, is seldom the case given the environmental conditions and price uncertainties which face farmers in many parts of the Third World. The lack of investment opportunities in economically depressed rural areas also helps to explain the predilection for consumption.

Such forms of expenditure do not only centre around satisfying households' basic needs, but may also take the form of 'conspicuous

Plate 5.2 'Saudi house': a common feature of villages in many parts of Thailand are the ornate dwellings which have been built with the proceeds of contract employment in the Arabian Gulf

consumption', whereby a migrant household may seek to enhance its actual or perceived standing in the community such as by constructing a large and elaborately decorated dwelling, or through the sponsoring of village festivals and ceremonies. In the Mossi region of western Africa, remittances from migration are commonly used for the giving of gifts, the accumulation of dowries and the acquisition of land, ahead of other potential uses, because of the considerable social importance of such things. Conspicuous forms of expenditure, such as on gold jewellery, modern clothes and various consumer items, may also be used by returning migrants to convey an impression of 'success' during their period away from the village. Quite often, however, migrants will use such conspicuous forms of expenditure to mask their lack of success. This in turn may lead to their conveying a false impression of what life and opportunities in the city are really like, and at the same time may raise the aspirations of others in the community, encouraging them to emulate their peers' 'achievements' through migration. In Liberia, return-migrants were considered responsible for contributing to higher expectations amongst younger (non-migrant) villagers, which may be expected to lead to higher levels of out-migration in the future.

Although often only of secondary importance, there are many instances where remittances have been used to good effect through their investment in various forms of productive enterprise. Investment expenditure may take the form of the purchase of farm machinery, various forms of farming inputs (fertilizer, pesticides, etc.), the undertaking of maintenance or improvement to the fabric of the farm, the purchase of land, and so on. In the Indian Punjab, formerly landless peasants have been able to acquire land through the income derived from migration, and others have been able to compensate for labour shortages by purchasing farm machinery. Similarly, in the north-east of Thailand the recent profusion of *khwaay lek* or iron buffalo (two-wheeled hand-tractors) has partly been facilitated by the financial injection which migration has given to village communities in the region. In the Turkish district of Bogazliyan, around half the tractors which were purchased between 1966 and 1975 were bought by migrants – not only for their utility value, it should be added, but also because of the prestige that possessing a tractor bestowed upon the owner.

Remittances may also have the effect of improving the distribution of income in source areas, although much depends upon which socio-economic groups are involved in migration and who is the more successful. If migration mainly involves people from the poorer echelons of village society, and if these migrants are successful in earning and saving a reasonable level of income from town, then migration may enhance their position relative to other groups. If, as seems more probable, migration streams include both the relatively wealthy and the relatively poor, then it is likely that the former will be more successful than the latter on account of their better education, better contacts and lesser inclination to take low-paid forms of employment, with the result that rural income disparities may be widened rather than reduced through migration. However undesirable this may be from the viewpoint of definitions of development which emphasize distributive justice, we should not lose sight of the importance of migration in providing a source of livelihood for households which might otherwise have few opportunities for economic advancement or, indeed, economic survival.

Not all transfers from city to village take the form of cash. Table 5.3 shows that migrants often bring a variety of consumer products back to their families when they visit or return permanently. Whilst it is difficult to quantify the total volume of such non-cash transfers, it is likely that

Table 5.3 Cebu (Philippines): non-cash remittances received and sent over a three-year period

Item	Gifts received in town Number	%	Gifts sent from town Number	%
Clothing	71	47.0	104	53.3
Food	26	17.2	52	26.7
Appliances	9	6.0	4	2.1
Watches, jewellery	5	3.3	2	1.0
Medicine	1	0.7	0	0.0
Combination (usually clothing and food)	28	18.5	25	12.8
Other	11	7.3	8	4.1
Total	151		195	

Source: Richard Ulack (1986) 'Ties to origin, remittances, and mobility: evidence from rural and urban areas in the Philippines', *Journal of Developing Areas* 18 (3).

they constitute a significant source of both basic and non-essential goods for migrant households. Table 5.3 also shows that reverse transfers between the village and the city are also of some importance. Thus, migrants may take food and other supplies with them to town when they leave their rural homes, not least because they are much more expensive to buy there. They may also take small sums of money with them to tide them over until they find work. Together with the transportation and other costs associated with migration, these 'reverse remittances' may be quite significant and may require that the migrant remains in town for some months before the household sees a return on its initial 'investment' (see Figure 5.1). Where people migrate to further their education, rather than to seek employment, there may occur a substantial net transfer of funds from the village to the city. Such is the case in western and northern Nigeria where large numbers of young people are migrating to Lagos and Ibadan to further their education.

Human qualities

Figure 5.1 suggests that there is a two-way interchange between the village and the city in respect of what we might loosely term 'human resources'. The selectivity of the migration process in general results in younger, better-educated, more dynamic and enterprising people moving away from their home communities – a form of 'brain drain'. The selective depopulation of rural areas may thus in turn be responsible for undermining the longer-term development potential of these areas by removing potential leaders and those who are arguably most receptive to change. On the other hand, the phenomenon of return-migration means that quite large numbers of people are coming back to their home communities, in most cases armed with the improved skills, experience and perspective that a period of life and employment away from the village may have provided. The fundamental question here is whether and to what extent the 'human qualities' that migrants bring back to their home areas compensate for those that are lost through migration.

The very fact that better-educated, more enterprising people are leaving rural areas suggests that they are unable to find employment and other opportunities which enable them to fully utilize their skills and potential, and thus their absence may not be too significant. In the short-term it makes economic sense for them to exploit their skills in the cities where such opportunities abound. However, this may become a self-perpetuating process. The steady cityward drift of the cream of the countryside is hindering the long-term changes and developments which are necessary to create the kinds of opportunities which would prevent them from moving in the first place. Furthermore, rural areas are shouldering an unfair burden in terms of the training, education and welfare of people who will ultimately join the urban workforce.

Although the better-educated and more highly-skilled rural migrants are unlikely to return permanently to their natal areas, there is evidence from a number of studies that the people who do return are generally more receptive to new ideas and more responsive to new opportunities. In Tanzania, and in the Highlands of Papua New Guinea, return-migrants were responsible for introducing new crops, such as cardamom and coffee respectively, and in Peru they were responsible for taking the lead with new horticultural innovations. In the Punjab, return-migrants were also instrumental in speeding up the introduction of improved, high-yielding crop varieties. The same was true of rural–rural migrants in Guatemala, who were responsible for the introduction of new crops and farming techniques in their home communities. Whilst migrants may thus be receptive to new ideas, this may also bring them into conflict with village elders and others who may be rather more conservative in their approaches.

Elsewhere, migration is argued to have played an important role in introducing a greater market orientation to agriculture, and also an expansion in non-agricultural activities in rural areas. Not only are migrants thus acting as catalysts of change, but they also possess the financial means, through their savings, to implement such changes. Migration studies in Malaysia and Thailand, on the other hand,

Case study E

The impact of migration in rural north-east Thailand: a mismatch of skills?

Roi-et is one of the poorest provinces in the poorest region of Thailand, the north-east. Agriculture provides the main source of livelihood for around 90 per cent of the province's population, and farmers have to contend with very poor sandy and saline soils, the regular threat of first drought and then floods, limited provision of irrigation, and isolation from the country's main market centres. With few jobs to occupy people outside farming, many thousands each year turn to Bangkok in the hope of supplementing their meagre incomes.

Because they have limited experience of work other than farming, most first-time migrants to the capital city find work in unskilled or semi-skilled occupations such as construction, taxi-driving, factory work and domestic service. A fortunate few may receive training or further their education whilst in Bangkok, and may then be able to gain access to better-paid and more secure jobs associated with tourism, commerce and industry. But the majority may expect to remain in the same forms of employment throughout their migration careers.

Those with the financial means to pay for work contracts and air fares may be able to work in the Middle East or in other parts of South-East Asia, where they can expect to earn considerably higher wages. Although the tasks they may undertake while abroad require few skills and qualifications, it is generally the better-educated and skilled workers who travel abroad. They may undergo a process of 'de-skilling', as they will have few opportunities to practise and develop their original trades and skills.

Most migrants from Roi-et feel a strong sense of attachment to their home areas, and the majority may be expected to return quite regularly during the course of migration careers which may span anything between five and twenty-five years in many cases. Having returned to their home villages, however, there is very little to keep them there. The incomes they can expect to earn from farming are very unattractive when compared with those available in the city. There is also very limited scope for them to

Case study E *(continued)*

Plate E.1 Migrant Thai construction worker in Brunei

utilize any skills they may have picked up whilst working in town. A few may be enterprising enough to set up their own construction or transportation businesses in their local areas, but the majority will quickly realize that a return to town offers the best prospects of using their skills and experience to the fullest effect.

In essence, then, there is a mismatch of skills between those offered by rural migrants and those needed in the city, and between those offered by return-migrants and the kinds of opportunities which are available in the countryside. Such a situation is clearly unsatisfactory from the viewpoint of the potential role that migration could play in supporting the development both of urban and rural areas. A more highly and appropriately skilled migrant workforce would greatly enhance the quality of the urban labour force. Similarly, the skills and experience potentially offered by return-migrants might make a significant contribution to the development of an economically backward, under-industrialized peripheral region.

Case study E *(continued)*

The Thai government has recently responded to this situation by establishing a number of skills training centres in the north-east which open a wider range of opportunities to city-bound migrants. In Bangkok, the multi-national firms which have played an important role in Thailand's recent rapid industrialization are required by the government to provide training for their unskilled and semi-skilled workforces. There are also a growing number of cases in the north-east where, with the assistance of the Department of Industrial Promotion, return-migrants have used the skills and experience they have derived from working in Bangkok factories to establish new, or modernize existing, small-scale rural cottage industries. The more successful of these enterprises have created a substantial amount of local employment, and have also stimulated the demand for local raw materials.

conclude that migrants are no more receptive to new ideas, new technology and other elements of change than non-migrants. Even those migrants who acquire particular skills whilst working away from the village may have very little scope to utilize them upon their return. Studies in rural parts of Columbia have shown that the work skills migrants bring back are of little relevance to the local economy. Similarly, rural Senegalese migrants returning from overseas work in France make little contribution to their local areas because the skills they bring back are wholly inappropriate to the kinds of economic activities undertaken in their home areas.

Thus, overall, it would appear that the rural areas from which the great majority of migrants originate in most Third World countries are something of a 'mixed bag' in terms of the impact of migration. Not only are there a variety of positive and negative effects, but both may be in evidence at the same time and within the same community. Similarly, what may be perceived by individual migrants to be a desirable outcome from migration may, in aggregate or collective terms, have strong negative connotations.

Whilst migration has provided the opportunity for a great many rural households to survive or even advance economically, it has also greatly increased the dependence of rural areas on the sources of livelihood

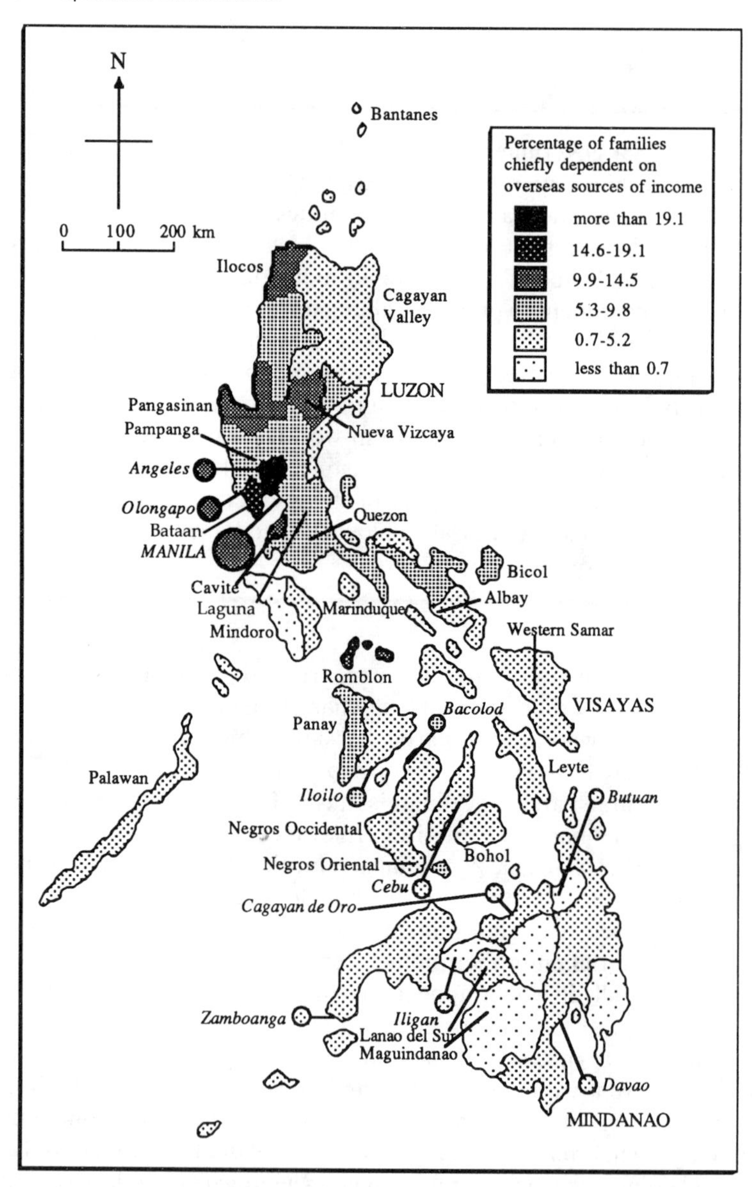

Figure 5.3 The Philippines: percentages of families which are dependent on overseas remittances for their main source of income, 1985

Source: R. T. Jackson (1990) 'The cheque's in the mail: the distribution of dependence on overseas sources of income in the Philippines', *Singapore Journal of Tropical Geography* II (2), p. 78.

that the city provides. A similar situation appertains in relation to the remittances sent by people working overseas (see Figure 5.3). It would be alarmist to suggest that this is an unhealthy dependence, but there have been instances where changing conditions in the city have affected rural migrants particularly badly. Once such situation has occurred in Mexico City over the last twenty or so years. Linkages with the city which were established through migration were largely responsible for the introduction of commercial livestock rearing in nearby rural areas for the Mexico City market, and the parallel decline in the production of food crops. By turning agricultural land over to pasture, farmers were able to increase their incomes whilst at the same time needing less labour (which was opportune given the scale of migration from these villages). However, increasing economic problems in Mexico City have resulted not only in a dwindling market for livestock but also a decline in seasonal employment opportunities for migrants from these rural communities. Their heavy dependence on migration and the urban market, and their neglect of food production, has caused significant problems in the rural source areas, and has subsequently resulted in a much higher level of permanent migration to the city and a decline in seasonal circulation which had hitherto been predominant.

Effects on destinations

In assessing the effects of migration on the places to which people move, we must again remember that the impact of migration depends upon who is moving, the circumstances which led to the move, the reasons for the migration, the characteristics of the movement, and the social, cultural and economic setting within which the migration is taking place.

One of the most obvious effects of migration is its contribution to urban growth in the Third World. As a rule of thumb, migration accounts for between one-third and one-half of the rate of Third World urbanization. In several Latin American countries, where the level of urbanization is already quite high and the rate of urbanization is slowing down, migration from the countryside to the city is relatively insignificant: even in the 1970s, rural–urban migration accounted for only around one-quarter of the total volume of movement in Chile and Peru, compared with migration between urban areas which constituted 40 per cent of migration in the former and 48 per cent in the latter. Much of this inter-urban movement may have consisted of step migration, and thus may initially have originated in the countryside. In countries such as Venezuela, Mexico, Brazil and Uruguay, perhaps the majority of movements now focus on intermediate cities, with the capital cities having lost their earlier attraction to migrants on account of rising levels of unemployment and government policies which have sought to redirect migration flows to towns and cities lower down the urban hierarchy. Only in Columbia has the trend towards a greater concentration of migration streams on the capital city continued.

Elsewhere in the Third World, there is still a pronounced tendency for migration to consist mainly of movements from rural to urban areas, and for the overwhelming majority of movements to be focused on the capital primate city. In Africa and Asia the rate of urbanization is also increasing rapidly. Here, in general, the larger the city, the greater is its rate of growth and the higher is the concentration of population migration on it. In Asia, rural–urban migration accounts for some 50 per cent of the rate of urbanization, with the remainder being made up by the natural increase of the urban population (which may itself reflect past rates of in-migration), and the reclassification of rural into urban areas. In many African countries the rate of internal migration is increasing quite rapidly.

We have also seen in earlier chapters that, in Asia and Africa, there is a significant volume of non-permanent migration to urban areas. Thus, whilst migration may be adding substantially to the rate of urban growth in these continents, the cities in fact play host to an even greater volume of movement than the rate of urban growth suggests. Therefore, at any one time the urban population may be made up of a significant number of people who originate elsewhere. For example, some two-thirds of the populations of the Ghanaian capital Accra and the Kenyan capital Nairobi are made up of people who were not born in these cities.

In general terms the effects of migration on the city are the reverse of those which we have just described for the village. Urban areas are a major recipient of labour from the countryside, much of it unskilled or semi-skilled, for which payment is made in the form of wages, part of which may be remitted back to the village. Urban areas may also contribute to the development of ‘human resources’ (training, education, experience) which may, to a limited extent, be of benefit to rural areas upon migrants’ return there. To a much greater degree, the city benefits considerably from the flow of labour from the countryside in that it provides the foundation for industrialization and other forms of economic development. Urban areas also benefit from the ‘brain drain’ effect of selectivity of the migration process, reaping the advantages of the qualifications, skills and enterprise of large numbers of rural migrants without having to make a significant investment in their education and training (and, indeed, welfare when unemployed). On the other hand, as has been described in other volumes in this series (see, for example, the books by David Drakakis-Smith, Janet Momsen, Alan Gilbert), several Third World cities have had to face up to the fact that labour (much of it derived from migration) is in over supply, a situation to which the large numbers of underemployed and swollen informal sectors clearly attest.

Not only have urban areas had to come to terms with an over-abundance of labour, but the large-scale influx of people from outside the city has also placed a massive burden on already over-stretched urban amenities and services, such as health, education, housing, water supply and sanitation, transportation and recreational facilities. The high cost of investment needed to deal with such problems often means that city authorities are unable to cope. It must be emphasized, however, that it is by no means clear that migration is the *major* contributor to such problems: we have seen in earlier chapters that a large proportion of migrants arrange employment, housing and other forms of support *before* they embark on their journey to the city. The main cause of such problems is the growing body of so-called 'urban poor', many but not all of whom may be migrants or descendants of people who earlier migrated to the city.

We should also be careful how we perceive the housing and other conditions with which migrants are faced. As we have seen earlier, most migrants consider themselves to be often much better off in town than in their place of origin. Furthermore, low quality housing often enables the migrant to save a greater proportion of his or her urban earnings than might otherwise be the case, which may be particularly important for short-term migrants who aim to maximize the return from their initial investment.

Whatever the relative merits or demerits of low-cost housing and a labour-absorbing tertiary sector, there is little doubt that in perhaps the majority of Third World countries the level of migration has tended to be higher than the capacity of the urban economy to absorb migrant labour. A possible explanation for the contribution of migration to 'over-urbanization' is the greater preponderance of 'push' factors which are forcing people into migration – courtesy of the rather dismal record of achievement with rural development – than 'pull' factors which attract people to real (as opposed to perceived) opportunities in the city.

Whatever the precise reasons, a significant body of the urban potential labour force (including, but not exclusively made up of, migrants) is underemployed for much of the time, which is clearly unsatisfactory both in terms of the welfare of migrant workers and in respect of the efficiency with which the urban labour force is being utilized. The oversupply of labour also has the effect of depressing urban wage levels which, whilst advantageous for industrialists and governments keen to underpin their competitive advantage over other countries, may not be such good news for the large numbers of migrants who rely on their urban wages to supplement their rural livelihoods.

Effects on individual migrants

For first-time migrants in particular, the move to town may be a very intimidating and alienating experience. We have seen that migration often involves young people who, for perhaps the first time in their lives, may have been given a considerable burden of responsibility in moving to town to supplement the livelihood of other family members. In addition they must come to terms with a move between two contrasting economic, social, cultural and physical environments. Table 5.4 provides a simple schema of some of the typical contrasts which may exist between rural and urban areas. In moving between them, migrants must seek also to adapt to life in an environment where the pace of life may be much faster than that with which they may be familiar, where individualism is rather more commonplace than communalism, where life focuses much more centrally around the workplace, and where cash provides the means of economic survival. In cases where migration takes place over considerable distances, the migrant may also encounter linguistic, religious and other cultural differences between place of origin and destination, and may be discriminated against by virtue of his or her discernible 'differentness' from the majority society. 'Street-wise' urban dwellers may appear to the new migrant to have a considerable advantage in terms of getting around, obtaining work and establishing social relationships, and may have developed ways of coping better with the sharp practices and bureaucracy of city people and institutions.

The new arrival from the countryside may be fascinated by the dynamism and excitement of the city, but more typically will feel bewildered, frightened, and perhaps also quite homesick (see Table 5.5). It is not unusual for people to feel like jumping on the next bus

Table 5.4 Some stereotypical characteristics of rural and urban areas in Third World countries

Rural	*Urban*
Rural setting: tranquil, serene	Hustle and bustle of urban life High-rise buildings; traffic
Caring social environment Central role of the family	Relationships more impersonal
Cohesive communities	Lack of a sense of community Discrimination against migrants?
Mutual support and assistance	Competition rather than cooperation?
Irregular burden of work Relaxed attitude towards work	Regular hours of work; regular wages
Work orientates around daily life	Life revolves around work
People produce their own food/goods	Basic necessities must be purchased
Limited need for cash Barter/exchange locally	People need cash to survive
Social sanction to pre-empt disputes	Few social controls on deviant behaviour

Table 5.5 Problems of adjustment to life in the city

Adjustment problem	*Never had any problems (%)*	*Used to have problems (%)*	*Still having problems (%)*	*Total (%)*
Male				
1 Housing	84.5	10.7	4.8	100
2 Work, place for further education or job training	77.7	15.1	7.2	100
3 Difficulty in getting along with people	82.4	14.4	3.2	100
4 Bad environment, crime, theft and gangsters	84.8	7.6	7.6	100
Female				
1 Housing	84.2	12.6	3.2	100
2 Work, place for further education or job training	81.6	13.5	4.9	100
3 Difficulty in getting along with people	83.4	14.6	2.0	100
4 Bad environment, crime, theft and gangsters	88.3	7.3	4.4	100

Source: Apichat Chamratrithirong, Krittaya Archavanitkul and Uraiwan Kanungsukkasem (1979) *Recent Migrants in Bangkok Metropolis: A Follow-Up Study of Migrants' Adjustment, Assimilation and Integration*, Bangkok: Mahidol University, Institute for Population and Social Research.

home. For most, however, this is not a realistic option because of the considerable investment which has gone into the migration, and the myriad hopes and expectations which may have accompanied the migrant to town. Also, the migrant is likely to be very reluctant to return to the village having failed to achieve his or her objectives: enduring the difficulties of life in town may be seen as preferable to the loss of face that may result from the migrant's failure to cope with life in the city. Thus for the majority of migrants an important need is to adapt as quickly as possible to life in the city.

Migrants can make life a little easier for themselves in a number of ways. One of the most commonplace methods of dealing with the transition between rural and urban environments is for migrants to familiarize themselves with the potential destination place(s) in advance of the migration, such as by visiting there or by seeking information from others who have previous experience of working there. In much the same way, as we have seen in earlier chapters, migrants are also very adept at mobilizing contacts in the city in order to pave the way for their arrival by arranging employment, accommodation and some friendly and familiar faces to help them through the most difficult immediate phase of their adaptation to life in the city. This explains why 'chain migration' is such an important phenomenon in many Third World countries. Migrants may also move in small groups who may help in the initial process of adaptation by providing mutual support. After arrival in the city the migrant may be expected to take up residence in an area in which a number of fellow residents from his or her home area may already be established. In the Philippines, slum communities which mirror the traditional character of rural settlements are important in providing a cultural 'staging post' for newly arrived migrants in the city. Mutual help provided by reconstituted family groups was also found to play an important part in the adaptation of migrants from the Dominican Republic to Venezuela.

There are also a wide variety of formal institutions which exist to help migrants adjust to life in a new environment. Migrant associations are generally organized by groups of migrants who have become established in a particular destination, and serve to assist others from their home areas who arrive there. These associations may be based around churches or sports clubs, as in Papua New Guinea where there are clubs representing different areas in all the main towns. The clubs may recreate elements of the home society in the city, thus helping to ease the transition between the two areas, and through the linkages which are maintained with the home area may facilitate migration in other ways, such as by providing information about employment opportunities. Migrant clubs and associations also play an important role in channelling money back to the home areas for various construction projects and festivals.

In Peru, a great many village communities are strongly represented in the capital Lima through various formal and informal village associations which have been founded by migrants. They fulfil many important social, recreational and welfare functions and also, through rotating credit systems and savings societies, help tide migrants over during financially difficult times. Migrant associations may also provide a foundation for political organization amongst the migrant community, enabling them to lobby governments on behalf of their community. In Mexico, village associations involving Mixtec (Indian) migrants to Mexico City have become heavily politicized, lobbying strongly for government assistance not only for Indian migrants in the city but also for the rights of Indian peoples as a whole in their home areas. Reflecting the greater difficulty that long-distance migrants have in adjusting to life in a new area, migrant associations are particularly prevalent amongst groups of people who come from areas which are quite far removed from the city, and are much less prevalent amongst migrants from nearby areas.

Whilst migrant associations and other mechanisms of adaptation may therefore help considerably in the process of adjustment to living in a new location, they may in fact slow the pace of assimilation – where the migrant becomes a part of the host society. For many migrants, particularly those who expect to remain in the city for only a relatively short period of time, this may not be seen as a problem: their social, cultural and emotional roots may remain firmly embedded in their home communities, and it may not seem worth while expending energy in trying to ‘become urban’. However, for people who intend remaining permanently in the city, it may be more important that they should ‘blend into their surroundings’, not least where people are discriminated against or looked down upon because of their rustic origins.

It does not necessarily follow, therefore, that the level of assimilation and integration are simply a function of the amount of time spent living in a particular place. There are many cases where people have lived away from their home areas for five or six decades and yet have not managed to pick up the language or dialect which is spoken in the destination area. Such a situation tends to be more noticeable amongst the older generations, who may be more set in their ways and who may find it rather difficult to change the habits of a lifetime. The younger generations may more readily become assimilated on account of their greater receptiveness to change, and also because they may have more opportunity to come into contact with members of the host society at school or in the workplace. The demands of education and employment make it imperative that they should be able to communicate and interact freely with members of the host society.

A number of migration studies have highlighted the ways in which migrants are perceived to change as a result of spending a period of time away from their home communities (see Table 5.6). Notwithstanding the subjective way in which such changes are viewed, the fact that migrants who may have spent only a relatively short time working in an urban centre are perceived to undergo a number of changes provides further evidence of the ways migrants not only adjust to but also assimilate into the host society. It is interesting to note that the majority of migrants are perceived to change in a positive way, although this may have much to do with the generally positive image of the city which pervades rural areas.

Table 5.6 Changes in return-migrants in north-east Thai villages as perceived by others in their home communities

Perceived changes	*Number*	*%*
More knowledgeable, experienced, clever	88	41.3
Improved appearance (clothes, cleanliness)	45	21.2
Improved behaviour	14	6.6
More mature and responsible	13	6.1
More progressive ideas	16	7.5
More extravagant, spendthrift	4	1.9
Worse behaviour	11	5.2
Worse appearance	2	0.9
Negative thoughts and ideas	2	0.9
Other changes	2	0.9
No perceived changes	16	7.5
Total	213	100.0

Source: The author.

Conclusion

It should be clear from the foregoing discussion that it is not possible to make a categorical statement as to whether or not migration is on balance desirable or undesirable from the perspective of development in Third World countries. We have seen that there are both positive and negative effects associated with migration, and that the migration process to a large extent centres around trading-off the costs against the benefits. These costs and benefits, or more generally 'better' and 'worse' developmental effects, are in any case very subjectively defined.

At the level of the individual migrant or migrant household, we might conclude that the overall impact of migration is beneficial. Migration would not continue in such massive volume if people did not perceive their prospects to be better through migration and if many were not indeed benefiting from migration. Whilst beneficial to the migrant, migration may be less so to others in their home and host communities. The often substantial cash infusions which are injected into the former may have an inflationary effect on the local economy, pushing up land and property prices beyond the reach of many. Urban dwellers, too, must suffer alongside migrants the congestion, pollution and competition for jobs to which migration makes its own small contribution.

At the level of the village, migration provides an answer but not a *solution* to the myriad problems which were outlined in Chapter 4. Provided the city continues to offer opportunities to migrants, and thus a potential escape route from rural poverty, the growing reliance of rural communities on the urban economy should not cause undue concern. If, as seems quite probable, the longer-term development prospects of rural areas are being undermined by the steady, highly selective drift of people towards the major cities of the Third World, the prognosis becomes rather less optimistic. Reliance becomes an unhealthy dependence, particularly in view of the very limited capacity of urban centres in many Third World countries to cope with a further increase in the size of the migrant population.

But what of the unevenness of the development process at the macro-level which, it was argued in Chapter 4, underpins a significant amount of migration in Third World countries today? To what extent is migration helping to redress the balance between rural and urban areas, and between the agricultural and industrial sectors? Migration cannot be viewed, as it was in the 1960s, as an equilibrating mechanism which is responsible for reducing these spatial and sectoral imbalances. Not only may migration be undermining the development prospects of rural areas, but it may also be making a significant contribution to the dynamism and comparative advantage of the major urban centres of the Third World. The city enjoys the benefits of a substantial inflow of semi-skilled and quite well-educated labour which, because it is in oversupply, can be employed very cheaply.

Migration therefore makes a significant contribution to urban-centred industrial development which, as we have seen earlier in this volume, provides the main engine of economic growth in many Third World countries and is also the main focus of government policies. Without migration it is doubtful whether some of the most rapidly-industrializing countries of East Asia and Latin America could have achieved rates of economic growth which would come close to those attained over the last decade or so. We cannot be so categorical about the positive effects that migration has had on the rural economy: agriculture has often languished as a result of the drain on human and other resources which migration implies. Non-farm activities, especially in rural cottage industries, also appear to have been weakened rather than strengthened through migration.

Thus, in assessing the effects of migration in Third World countries, scale becomes a very important factor. A great many people benefit from migration, if only through escaping deprivation and exploitation. And yet at the macro-scale migration would appear to be doing very little to reduce, and much to widen, the economic and social disparities which give rise to a significant proportion of movements in the first place. From the viewpoint of national economic development such an outcome may not be wholly undesirable because it perpetuates an environment conducive to urban-industrial centred economic growth. But in terms of social justice, especially in rural areas, an outcome which sees migration as the principal solution to rural development problems is both unfortunate and unfair.

Key ideas

1 Even during slack periods in the farming calendar, rural labour is seldom redundant. Thus, even seasonal forms of circulation will have an effect upon rural production, particularly in relation to off-farm activities.
2 The tendency for consumption to be given a higher priority than investment in the use of migrants' remittances is entirely logical in the face of the environmental and economic uncertainties which face farmers in many parts of the Third World.
3 The institutions and mechanisms which exist to ease a migrant's entry into the host society may speed up the process of adjustment but,

because they tend to draw migrants closer together, may slow the pace of their integration.

4. There is a contradiction between the micro-level effects of migration, where the majority of migrants perceive themselves to be better-off as a result of their movement, and the macro-level effects, which generally point to the disadvantaging of rural source areas and a mixture of positive and negative effects on the main urban destinations.
5. Migration may be responsible for perpetuating, or even widening, the various forms of uneven development to which it is also largely a response.

6
Population movements in the Third World: policy and planning issues

A United Nations survey in 1978 found that 116 out of the 122 Third World countries which were surveyed had devised policies to decelerate the rate of rural–urban migration and to influence the spatial distribution of population. Clearly, however much individual migrants may consider themselves to be better-off as a result of their move, their respective governments are concerned about the longer-term developmental implications of the drift of people from the countryside and their subsequent concentration in urban areas.

The design and implementation of policies aimed at influencing the scale and pattern of population movement is an onerous task. We have seen in earlier chapters that the planners are faced with a mind-boggling array of movement types, each of which occurs in response to a wide variety of stimuli and leads to a complex range of positive, negative and neutral effects. Thus blanket policies which treat movements and movers as homogeneous entities are unlikely to be effective in anything more than a fraction of cases. Policies which seek to restrict movement by erecting barriers to personal mobility not only impinge upon people's rights and freedom but are also guilty of tackling the symptoms of the problems which give rise to migration instead of the root causes. There are also too many gaps in our knowledge and understanding of the migration process, and in the data which measure it, for planners to be able to make clear and unequivocal decisions as to how to plan for migration in the future. Finally, it is far from certain that planners will be able to come up with workable 'solutions' to the problems associated with population movement: after all, is it not the shortcomings of the planning system in adequately coming to terms with the phenomenon of uneven development which are largely responsible for the incidence of migration in the first place?

The ideal objective of migration policies should be to maximize the positive effects and to minimize its negative impact. This in itself is no

Table 6.1 A typology of migration policy approaches

Policy approaches	*Rationale*	*Types of policies and programmes*
Negative	Emphasizes the undesirability of migration and seeks to erect barriers to population movement and to forcibly 'deport' migrants	Closed city; pass laws; deportation of beggars, the homeless and those in marginal occupations; bull-dozing of squatter settlements; enforced resettlement from urban to rural areas; sedentarization of nomads; registration systems; employment controls; restricted access to housing; food rationing systems; benign neglect.
Accommodative	Accepts migration as inevitable, and seeks to minimize the negative effects in both origin and destination places	Slum up-grading; sites and services; urban job creation – labour-intensive industrialization; minimum wage legislation; urban skills training; urban infra-structural investment; improved social welfare; improvements in transportation; relieving congestion.
Manipulative	Accepts migration as inevitable and even desirable in some cases but seeks to redirect migration flows towards alternative destinations	Colonization; land settlement; land development; polarization reversal; growth poles; urban, industrial and administrative decentralization; information systems; management of contact networks.
Preventive	Rather than dealing with the symptoms of migration, attempts to confront the root causes by tackling poverty, inequality and unemployment at source, and reducing the attractiveness of urban areas to potential migrants	Land reform; agricultural intensification; agricultural extension; rural infrastructural investment; rural industrialization; rural minimum wages legislation; rural job creation; improving rural–urban terms of trade; reducing urban bias; increased emphasis on 'bottom-up' planning; propaganda in favour of the rural sector.

easy task, given the difficulty we have in deciding what is 'positive' and 'negative', and also given the fact that both outcomes may result from the same form of movement. The following discussion outlines some of the main policy responses to migration and other forms of population movement. In general these have moved away from the draconian preventive measures which were widely practised in the 1960s and early 1970s towards the rather more realistic and humane policies which are widespread today and which accept that migration is inevitable given the current characteristics of development in Third World countries.

The main forms of policy approaches to influencing the incidence and impact of Third World population movements are summarized in Table 6.1. This simple typology identifies four broad forms of policy, each of which will be discussed in the following sections.

Negative approaches

Policies which dwell on the negative effects of migration and which attempt to deal with these by preventing the movement of population are fortunately largely a thing of the past. These were mostly unenlightened policies implemented by authoritarian governments concerned with the perceived effects of the large-scale drift of people towards the cities rather than with the welfare of the migrants or the conditions which were giving rise to their movement. The realization that the erecting of barriers to movement deals with the symptoms of a variety of development problems, rather than with the problems themselves, has in recent years caused policy-makers to seek to work more on behalf of migrants than against them. None the less, several Third World governments continue to make life extremely difficult for migrants in the hope that many will eventually return to their rural homes.

In the majority of cases negative policy approaches highlight the authorities' lack of understanding of, or concern for, the factors which cause people to leave their home areas. During the Vietnam War, the authorities in Saigon (now Ho Chi Minh City) sought to counter the large-scale movement of people towards the capital city of South Vietnam with a policy of benign neglect, believing that the more intolerable conditions became in the city, the less inclined people would be to move in and the more inclined those already there would be to leave. Although conditions in Saigon did indeed become quite awful, the 'policy' disregarded the severe disruption which was then taking place in the countryside and which was more or less forcing people to seek refuge in the city. As a consequence, the policy led to considerable misery and hardship for the city's inhabitants without leading to a substantial decline in the size of Saigon's population.

A similarly unenlightened policy, born in part out of desperation at the massive level of in-migration, was implemented in Indonesia where, in 1970, the capital Jakarta was declared a 'closed city'. To be allowed to enter the city migrants had first to get permission to leave their home area and obtain a 'short visit' card. They were also required to deposit with the authorities a sum equivalent to two times the fare to Jakarta, which presented an almost insurmountable barrier to migration amongst the poorest of the poor (as the authorities had intended). If within six months of arriving in the city migrants could provide proof of having obtained work and accommodation, their deposits would be refunded and they would also be given an identity card proclaiming them to be citizens of Jakarta. Migrants who could not comply with these regulations were 'deported' from the city.

In addition to this policy, the authorities in Jakarta also cracked down strongly on street vendors and trishaw riders, many of them migrants, who were operating without licences. A similar programme was implemented in the Peruvian capital Lima in the early 1980s. Operatives were systematically excluded from the more lucrative city centre locations, and instead had to scratch a living in more marginal peripheral zones. The policy was also effective in identifying migrants who had managed to circumvent the city's migration controls, but in general it did little to ease the pressure of migration on the city's over-burdened infrastructure because, once again, it failed to tackle the root causes of migration to the city, many of which lay in the underdeveloped state of the rural sector in Java. It also added further to the misery of Jakarta's migrant population, especially those people who were already operating at the very margins of an urban existence. Although the authorities in Jakarta, and other large urban centres in Indonesia, continue to try to restrict in-migration through the identity card system, since the 1980s the Indonesian government has accepted the inevitability of migration in the longer-term, and has recently adopted more humane policies which plan to accommodate future urban growth resulting from migration.

Other, even more radical programmes have been implemented in the Communist world. The government of Fidel Castro has, since 1964, practised a 'return to the land' movement in Cuba which has seen the population of the capital Havana remain more or less constant over the

Case study F

Migration controls in the People's Republic of China: measures to restrain urban growth

Before coming to power, the Chinese Communist Party had envisaged a rapidly urbanizing and industrializing China. Mao Tse-tung anticipated that tens of millions of rural peasants would migrate towards the cities where they would help swell the country's industrial workforce.

After 1949, improving conditions in urban areas, coupled with widespread difficulties in the rural sector, did indeed make the larger cities in particular highly attractive to migrants from the countryside. Accordingly, the cities grew very rapidly – so rapidly, in fact, that the authorities quickly became concerned about rising levels of unemployment and deteriorating housing conditions among certain segments of the urban population. In spite of these difficulties, income and welfare differentials between rural and urban areas also began to increase rapidly, which contradicted the principles which had underpinned the Communist revolution.

In response, the Party instituted a variety of measures designed to slow the drift of population towards the cities and, at the same time, several more stringent measures aimed at transferring large numbers of urban dwellers to the countryside. From the 1950s onwards, attempts to control the rate of cityward migration centred around the use of a population registration system which identified people either as 'urban' or 'rural'. People had to obtain permission to leave the countryside, and they could only move to work in the city if they could produce documentary evidence that they had a job to go to. By limiting the numbers of citizens who could be registered as urban dwellers, people were restricted from moving opportunistically to the cities in the hope of finding work, which has provided the impetus behind urbanization in most capitalist developing countries.

In addition to the registration system, the authorities also used food rationing to restrict the scale of rural–urban migration. Grain and oil rations in the cities were made available only to people in possession of urban household registration documents. With the limited operation of a black market, the rationing system

Case study F *(continued)*

proved a very effective mechanism for slowing the rate of rural emigration.

Since the 1950s, the Chinese authorities have also periodically encouraged large numbers of people to leave the cities, some voluntarily, others very reluctantly, as a way of alleviating the mounting problems in urban areas and also as a way of spreading development to the countryside. In the 1950s and 1960s large numbers of people were sent from the cities to facilitate the development of oil-fields in northern and north-eastern China, and to colonize virgin land for the cultivation of cereals.

There have also been series of 'sendings down' (*xia fang*) of urban residents to the villages. The 'back to the villages' (*huixiang*) movement in the early 1960s resulted in the movement of some 20 million people from the country's major cities, many of whom had earlier moved towards the cities as a result of the rapid industrialization which took place during China's 'Great Leap Forward' in the late 1950s. There were also mass deporta-

Table F.1 Official and unofficial levels of urbanization in the People's Republic of China, selected years between 1949 and 1988 (millions)

	Official Figures		*Unofficial Figures*		
Year	*Urban population*	*per cent of total*	*Urban population*	*per cent of total*	*Total population*
1949	57.65	10.6	49.00	9.1	541.17
1953	78.26	13.3	64.64	11.0	587.17
1957	99.49	15.4	82.18	12.7	646.53
1960	130.73	19.7	109.55	16.5	662.07
1964	129.50	18.4	98.85	14.0	704.99
1969	141.17	17.5	100.65	12.5	806.71
1976	163.41	17.4	113.42	12.1	937.17
1980	191.40	19.4	140.28	14.2	987.05
1984	331.36	31.9	174.42	16.8	1,038.76
1986	441.03	41.4	200.90	18.9	1,065.29
1988	550.00	50.0	230.05	20.9	1,100.00

Source: Terry Cannon and Alan Jenkins (eds) (1990) *The Geography of Contemporary China: The Impact of Deng Xiaoping's Decade*, London: Routledge, p. 210.
Note: The official urbanization figures include a large number of rural people who live within 'municipal' boundaries. For the purposes of the present discussion, the unofficial figures may be considered to be the more representative.

Case study F *(continued)*

tions of urban youth to the countryside from the mid-1950s onwards. These mass 'sendings down' not only helped to relieve population pressure in the cities and defuse discontent amongst unemployed youths, but were also designed to help raise rural productivity by sending well-educated youths out to the villages. The programme continued until the late 1970s, whereafter it was put into sharp reverse in support of China's industrialization strategy. By 1983 the majority of young people who had been sent down to the villages had returned to their native towns and cities. Whilst the sending down programme may have been successful in slowing the rate of urban growth, albeit temporarily (See Table F.1), it has not been without problems. Rural peasants have often resented the large-scale influx of strangers from urban areas, whom they were supposed to educate in the ways of the countryside. There was also considerable resistance to the sending down programme in the cities, where living conditions continued to be much better than in rural areas.

In contrast to the situation in many free-market economies, in-migration accounted for only around 30 per cent of urban growth in China during the course of the 1950s, 1960s and 1970s. Shanty towns, underemployment, poverty and begging also appear to be much less prevalent in China's cities than is the case in other Third World countries, although this has been achieved at a cost of greatly restricted personal freedom.

During the 1980s China has relaxed many of its controls on rural–urban migration as part of its push towards rapid industrialization. Table F.1 shows that this has resulted in a dramatic increase in the rate of urban growth. It remains to be seen whether this shift in policy will lead to many of the problems which have tended to be associated with rapid urbanization elsewhere in the Third World.

Source: R. J. R. Kirkby (1985) *Urbanization in China: Town and Country in a Developing Economy, 1949–2000 A.D.*, London: Croom Helm.

last three decades by restricting people's need and freedom to move to there. This programme has been paralleled by the 'urbanization of the countryside', which has aimed to create a spatially and demographically more balanced distribution of population. China has similarly attempted to restrict urban growth whilst at the same time seeking to underpin the development of the rural sector. During its famous rustication programme, an estimated 10 to 15 million urban school leavers were resettled in rural areas between 1969 and 1973. The policy aimed simultaneously to ease the burden on employment opportunities in the cities and to redirect better-educated people to the countryside where they could help to raise levels of rural productivity. More recently, the Chinese authorities have also used restrictive measures such as pass laws, which govern people's access to food rations and accommodation, to control the drift of rural people to the country's major urban centres (see Case study F).

Both sets of policies have to be viewed against a backdrop of broader programmes of reform which have sought to fundamentally change the pattern and process of development within their respective territories, and which have severely curtailed people's rights of freedom of movement. None the less, in terms of controlling a number of the problems which have come to be associated with rapid urbanization in most free-market countries in the Third World, both policies can claim a considerable degree of success. Where restrictive migration policies are implemented outside the framework of such radical reform programmes they have tended to remove the 'pressure valve' effect of migration without offering much in the way of alternatives. They also deny migrants access to the economic and welfare improvements which many perceive they enjoy as a result of migration.

Whatever the arguments against policies which restrict population movements, they have in general foundered on logistical rather than humanitarian grounds. It is extremely difficult to implement policies which seek to control the volume of movement from rural to urban areas. It is impractical to erect check-posts on all major roads leading to the cities, and it is also very difficult to enforce pass laws effectively. As with the Indonesian example presented earlier, widespread negative policies are often symptomatic of a government's inability to come to grips with the factors which are leading to migration in the first place. Whilst these persist, and whilst the cities continue to offer the prospect of a better life, people will continue to run the gauntlet of arrest and imprisonment in the hope of availing themselves of these opportunities. Not until the causes of migration are adequately confronted can policy-makers expect to see an easing in the momentum of cityward migration, whatever barriers they may wish to erect in its path.

Accommodative approaches

The three remaining types of policy approach have an implicit acceptance of the inevitability of migration, but differ in the ways in which they deal with it. The accommodative approach is principally concerned to minimize the negative effects of population movements both in the areas from which migrants originate and the major places of destination. It dwells upon the positive attributes of migration, such as its potential contribution to industrialization, its role in redistributing income from urban to rural areas, and its contribution to the development of urban–rural linkages.

In the cities, policies may focus on improving the welfare of the migrant population, such as through introducing minimum wage legislation, regulations concerning working conditions, and through job-creation programmes which may focus on encouraging labour-intensive production techniques. The Malaysian government helps hawkers in Kuala Lumpur and other urban centres by providing training and capital and marketing assistance, and thus seeks to remove some of the constraints which place them in a cycle of poverty. Several Third World countries now have policies which legitimize squatter settlements and shanty towns, and which thus remove the threat of eviction which may hitherto have added to dwellers' lack of security. Although these policies might typically be targeted at the 'urban poor' in general, rather than migrants in particular, there is little doubt that they are of benefit to the substantial section of the migrant population which ends up having to eke out a living in the city.

In rural areas policies may focus on the education and training of potential migrants. Whilst this may actually encourage people to leave their home areas, as they have marketable skills to sell, at the same time it helps to prevent a situation where migrants are forced into low-paid, marginal forms of employment in the city because of their lack of skills and qualifications.

The accommodative approach does not, in itself, represent a long-term solution to the problems associated with migration because it, too deals with the symptoms rather than the causes. None the less, the approach is infinitely more humane than that discussed above in that it recognizes people's need and right to move to the city, and seeks to ameliorate the hardship that they may experience there. It has been criticized, however, because it does little if anything to dissuade other people from moving towards the cities, and thus may contribute to a worsening of problems in the longer term which the city authorities may not be able to cope with.

Manipulative approaches

Manipulative policies also accept migration as inevitable and not altogether undesirable but, instead of seeking to ameliorate the negative consequences, focus on changing the pattern of population movements so as to spread potential benefits more widely. In simple terms, they are concerned to encourage the movement of employment and other economic opportunities to where they are needed rather than relying on people having to move to avail themselves of these opportunities. Manipulative policies are underpinned by the belief that the present pattern of migration, which predominantly consists of movements of rural dwellers towards the major urban centres of the Third World, is undesirable both because of 'overurbanization' and because it selectively drains the human resources of peripheral rural areas. The policies focus on redirecting the flow of population towards alternative destinations, in particular the smaller intermediate cities which lie further down the urban hierarchy. The resultant volume of short-distance local movements will not only support the development of activities in smaller provincial and regional centres and strengthen linkages between these towns and their rural hinterlands, but may also help to ease the burden on the major metropolitan centres of the Third World.

Thus the manipulative approach maintains people's freedom of movement whilst governments seek to influence people's choice as to where to move. We have seen in earlier chapters that people's perceptions of opportunities, and their personal contacts, play an important role in influencing their choice of destination. Manipulative policies therefore focus not only on creating opportunities in a wider range of urban areas but also on informing potential migrants about their existence and about how to avail themselves of them.

Urban-industrial decentralization and the creation of regional growth poles provided the cornerstone of manipulative migration policies during the 1970s and early 1980s. In some countries (e.g. Brazil and Pakistan), new capitals were established, with the help of considerable government investment, as a way of creating alternative focuses for cityward migrants. In Venezuela, planners used the country's revenue from petroleum to create a new town of some 300,000 people, Ciudad Guyana, partly as a means of spreading economic activity to the relatively impoverished south-east of the country but also to act as a counterweight to Caracas as an attraction to migrants. In some respects the policy was too effective because migrants flocked to the city in much greater numbers than could be adequately catered for. In the early 1970s there was a deficit of around 46 per cent in the city's housing stock, leading to the development of shanties. The massive rate of in-migration was also reflected in a severely unbalanced age structure of the population, with a disproportionately high number of young adults.

The government of Thailand has also sought to reduce the very pronounced emphasis of migration flows on Bangkok by encouraging industrial firms and public sector departments to relocate to designated regional centres, in the process creating local opportunities which might deter potential migrants from moving towards the capital city. Unfortunately, very few firms have been willing to forego the advantages of location in close proximity to Bangkok in exchange for the government's very generous fiscal incentives to move to the peripheral regions. Thus, very few opportunities have been created which might serve to deflect migration streams towards regional urban centres. Indeed, there has even been a suggestion that the rate of migration towards Bangkok may actually have increased, particularly amongst the better-educated, as a result of other improvements which have taken place in the peripheral regions due to the government's decentralization policies, including considerable investment in education, training and communications.

The shortcomings of decentralization policies in influencing the pattern and rate of migration have led to substantial modifications during the 1980s and 1990s. Most governments now accept the difficulties which are encountered in encouraging factories and migrants to move towards intermediate cities in peripheral regions, and have opted instead to promote shorter-distance movements to satellite towns around the main urban centres. Such a shift in policy has been very successful in Mexico, and in South Korea where a considerable volume of migration has been deflected away from the capital Seoul. Unfortunately, whilst such satellite policies may have been effective in influencing migration patterns around the major urban centres of the Third World, they none the less perpetuate the concentration of migration streams on the economically more dynamic capital regions. Thus they are only partially achieving their objectives of spreading the pattern of economic activity, and with it patterns of migration, more widely. So it is the poor design and implementation of these manipulative policies, rather than the policies themselves, which are responsible for their limited success in influencing patterns of population movement.

Preventive approaches

None of the policy approaches which we have examined thus far gets to grips with the root causes of migration, particularly the conditions which appertain in rural areas to which a large majority of migrants in the Third World respond. Yet it is here that any long-term solution to the problems associated with rural–urban migration must surely be based. The final category of approaches, preventive measures, is concerned with policies which deal not so much with the symptoms but with the affliction itself; policies which confront the reasons for migration, such as poverty, inequality and unemployment at source, and which restrict the attractiveness of urban areas to potential migrants. Although the results may be slow and unspectacular, there is little doubt that, in the longer-term, they may be more effective in influencing both the rate and pattern of population movements in Third World countries than has been the case with the reactive policies described above. As we shall see, however, the inappropriate nature of many such policies, and their poor implementation, has often restricted their effectiveness in terms of influencing the incidence and pattern of population movements.

The discussion in Chapter 4 suggested that land, employment and income are among the most important factors which influence the migration decisions of people in rural areas. It thus follows that policies which attempt to tackle problems such as land maldistribution, under-employment and poverty in the rural sector may, in some but not all cases, be effective in reducing people's need to migrate. A number of Third World countries have introduced programmes of land reform aimed at redistributing land resources and giving the landless and land-hungry the chance to earn a reasonable livelihood from farming. Whilst potentially removing the need for people to leave rural areas because of the inadequacy of their land-holdings, or because of exploitation by large land-owners, the record of achievement in this regard is rather disappointing because few Third World governments possess the political will or ability to successfully implement their land reform programmes.

It is also far from clear that the successful programmes have been effective in reducing the rate of out-migration. The break-up of sugar plantations in the coastal regions of Peru in the late 1960s and early 1970s, and the establishment of agricultural cooperatives, was very effective in terms of raising levels of production and productivity, but this was achieved at the cost of increasing rates of out-migration as surplus labour was freed from the land as levels of efficiency rose. A similar situation has occurred in other Latin American countries where the dismantling of feudal farming systems through land reform has led to increased efficiency and a parallel displacement of large numbers of rural inhabitants to find work in other areas.

Agricultural intensification programmes have also been rather ineffective in reducing the flow of population from rural areas, in spite of their potential for creating conditions which might allow more people to be supported by land-based occupations. In most cases food security, rather than increased employment opportunities, has provided the main motivation for such programmes. The Muda Scheme in north-west Malaysia sought to increase wet-rice production, and thus reduce the country's rice exports, by investing in a major irrigation scheme in the main rice-producing region. Although the scheme was very successful in raising levels of production and productivity, employment opportunities for landless labourers actually declined, in spite of growing two rice crops each year, because of the widespread tendency to use machinery in preference to labour. Former tenant farmers were also displaced from the land because farming became so profitable that landowners claimed back their land so they could work it themselves. Many of the former labourers and tenants had no choice but to seek work in local towns.

Thus, whilst they might potentially prevent migration by productively absorbing rural workers *in situ*, these policies have been largely ineffective because of the understandable priority they afford to engendering economic efficiency. Reducing the rate of out-migration is very often a hoped-for result rather than an explicit policy objective. Rural industrialization, on the other hand, is often put forward as a strategy for retaining people in rural areas whilst at the same time underpinning the development and diversification of the rural sector. The Thai

Plate 6.1 Rural industrialization. The manufacture of industrial products in rural areas offers the prospect of a genuine form of decentralized development which may help to deflect a considerable volume of migration in many Third World countries

government has recently afforded a high priority to the 'industrialization of the countryside' through which it aims to reduce people's need to migrate by creating more and varied income-earning opportunities in villages throughout the country, and especially in the peripheral northern and north-eastern regions from whence the majority of city-ward migrants originate. The policy focuses on the modernization of cottage industries and the establishment of putting-out systems where rural households become involved in parts of the manufacturing process for urban-based industries. Although it is rather early to judge the success of the programme, it appears that the more dynamic rural industries have not only succeeded in reducing the drift of young people towards the cities, and Bangkok in particular, but have also attracted migrants from other areas. The Thai experience contrasts with that of India, where attempts to develop small-scale cottage industries sometimes led to an increase in out-migration because it improved villagers' skills, and thus also their prospects of finding work in urban areas.

The government of Sri Lanka has been quite successful in slowing the rate of rural–urban migration by reducing the welfare gap between rural and urban areas. This it has achieved by improving medical services and education facilities in the countryside, through providing income support in the form of guaranteed prices for farmers, and by improving the quality of rural housing, especially for lower-income groups. This integrated package of policies has helped to reduce people's incentive to migrate towards Colombo and other urban centres by tackling another of the factors which influences the migration decisions of rural folk: the perceived welfare differential between the village and the city. Because image and perception are important in influencing both the incidence and pattern of movement, the government has also focused on promoting propaganda in favour of the countryside as a way of reducing people's inclination to move.

Conclusion

This chapter has identified a number of policy approaches to dealing with the problems associated with population movement in Third World countries. In the main, they have been rather ineffective in terms of slowing the rate of cityward migration, ameliorating the negative consequences of migration and promoting its positive effects. There are a number of reasons for these shortcomings in planning policy. First, policy-makers are often insufficiently aware of the full range of factors which influence migration decisions. Second, policies have often focused on 'single variable solutions' such as improving urban housing or creating rural jobs when, as we have seen, migration occurs in response to a complex variety of factors and stimuli. Third, influencing migration is often a hoped-for consequence of broader policies of rural and regional development instead of being a policy objective in its own right. Finally, migration-specific policies have generally been implemented against a backdrop of broader development programmes which continue to emphasize rapid economic growth which is predominantly city and industry focused. This in turn is having the effect of heightening the conditions of uneven development which, it was argued earlier in this volume, itself underpins a considerable volume of movement in Third World countries.

The countries which have been most successful in influencing the pattern and level of movement have either been those which possess the financial resources to adequately fund their various development programmes (e.g. Venezuela, South Korea, Malaysia), or which have the political commitment and authority (as in the case of China and Cuba) to effectively implement their migration policies. Given that the large majority of Third World governments are not blessed with the resources, authority and will to influence the pattern and process of migration, it seems inevitable that the future will see a continuation of the *ad hoc*, reactive policies which have hitherto been most prevalent. What this means to the peoples of the Third World is that they may expect to continue to gain access to economic and other opportunities through migration, but that in the process they may continue to undermine the prospects for a self-reliant improvement in conditions and opportunities in their home areas in the longer term. Market forces, and not government policy, will continue to determine who is able to move, who benefits, and who suffers as a result of migration.

Key ideas

1 Policies which erect legal and institutional barriers to the movement of population in preference to confronting the root causes of migration lack compassion and foresight, and do not offer a long-term solution to the cityward drift of rural people.
2 The ideal objective of migration policies should be to minimize their negative impact whilst at the same time maximizing the positive effects.
3 Policies which have sought to restrict rural–urban migration by improving conditions in migrants' areas of origin have very often had the opposite effect by displacing people from the land in the drive for greater economic efficiency, and also by raising skills, needs and expectations beyond the capacity of the local area to satisfy.
4 The more the authorities try to ameliorate the difficult conditions that face migrants in many of the Third World's largest urban centres, the greater the danger is that they may increase the attractiveness of these cities to potential migrants in rural areas.
5 The governments which have been most successful in influencing the pattern and volume of migration are those which possess the financial resources, the authority and the political will to effectively implement their policies, and those which have successfully integrated migration policies into their broader programmes of economic development and social change.

Review questions and further reading

Chapter 1

Review questions

1 Why is mobility so important to the peoples of the Third World? Why are some people more mobile than others?
2 If migration can be seen as the 'litmus test' for development, what do contemporary patterns of movement tell us about the characteristics of development in the Third World?
3 List some of the more important roles that population movements have played in the history of humankind.
4 If migration for many represents a 'survival strategy' and 'pressure valve' for the problems of source areas, what would be the likely consequences of restricting people's freedom to migrate?
5 Why is there sometimes a wide difference between migrants' expectations before their move and the realities of life in their chosen destinations?

Further reading

Jackson, John A. (1986) *Migration*, Harlow: Longman.
Kosinski, Leszek A. and R. Mansell Prothero (eds) (1975) *People on the Move: Studies on Internal Migration*, London: Methuen.
Lewis, G. J. (1982) *Human Migration: A Geographical Perspective*, London: Croom Helm.

Chapter 2

Review questions

1 Why should there be a link between the ways that we define population movements and the volume of movement which is identified?
2 Using the various bases of classification outlined in Chapter 2, calculate the number of permutations of movement types which might potentially occur. On the basis of the discussion in this volume, list the more prevalent of these movements which occur in the Third World setting.
3 What do you consider should be the minimum amount of time a person should be away from their home area before their movement takes on some significance? Explain the reasons for your choice. What factors would you use to identify permanent migrants, bearing in mind that there always remains the prospect that they will return to their home area at some stage in the future?
4 How clear is the distinction between voluntary and involuntary population movements? The former may offer little realistic choice other than movement, whereas the latter often involve a certain degree of choice about whether or not to move.

Further reading

King, Russell (ed) (1986) *Return Migration and Regional Economic Problems*, London: Croom Helm.
Prothero, R. Mansell and Murray Chapman (eds) (1985) *Circulation in Third World Countries*, London: Routledge & Kegan Paul.
Shrestha, Nanda R. (1988) 'A structural perspective on labour migration in underdeveloped countries', *Progress in Human Geography* 12 (2), pp. 179–207.

Chapter 3

Review questions

1 Several of the forms of population movement which are described in Chapter 3 are not exclusive to the Third World. Identify some of the counterpart types of movement which may be found in First World countries. Where differences occur, how might these primarily be explained?
2 What are your views about the ways in which traditional forms of population movement, such as nomadism and transhumance, are being increasingly restricted by the encroachment and impact of commercial agriculture and modern economic systems?
3 Political asylum in First World countries is increasingly difficult to obtain as governments have become more and more sensitive about the issue of

immigration, and as economic migrants from Third World countries have sought to circumvent immigration restrictions by claiming refugee status. How would you draw the distinction between a genuine refugee and an economic opportunist?

4 Should 'guest workers' be allowed rights equal to the citizens of their host countries, and should they be allowed to remain in these countries now that the conditions which underpinned their original migration have, in the majority of cases, changed?

Further reading

Black, Richard (1991) 'Refugees and displaced persons: geographical perspectives and research directions', *Progress in Human Geography* 15 (3), pp. 281–98.

Castles, Stephen (1986) 'The guest-worker in Western Europe – an obituary', *International Migration Review* 20 (4), pp. 761–78.

Condon, Stephanie A. and Philip E. Ogden (1991) 'Afro-Caribbean migrants in France: employment, state policy and the migration process', *Transactions of the Institute of British Geographers* (new series) 16 (4), pp. 440–57.

Cross, M., and H. Entzinger (eds) (1988) *Lost Illusions: Caribbean Minorities in Britain and the Netherlands*, London: Routledge.

Eickelman, Dale F. and James Piscatori (1990) *Muslim Travellers: Pilgrimage, Migration and the Religious Imagination*, London: Routledge.

King, Russell (ed) (1985) 'European migration: the last ten years', *Geography* (special issue) 70 (2), pp. 151–82 (international migration from the Third World: pp. 151–68).

Portugali, Juyal (1989) 'Nomad labour: theory and practice in the Israeli–Palestinian case', *Transactions of the Institute of British Geographers* (new series) 14 (3), pp. 207–20.

Simon, David (1989) 'Rural–urban interaction and development in Southern Africa: the implications of reduced labour migration', in Robert B. Potter and Tim Uwin (eds), *The Geography of Urban–Rural Interaction in Developing Countries: Essays for Alan B. Mountjoy*, London: Routledge, pp. 141–68.

Stahl, Charles W. and Fred Arnold (1986) 'Overseas workers' remittances in Asian development', *International Migration Review* 20 (4), pp. 899–928.

Swindell, K. (1979) 'Labour migration in underdeveloped countries: the case of sub-Saharan Africa', *Progress in Human Geography*, 3, 219–59.

Wilkinson, S. Robert (1983) 'Migration in Lesotho: some comparative aspects with particular reference to the role of women', *Geography* 68 (3), pp. 208–24.

Chapter 4

Review questions

1 Draw up a list of the positive, negative and neutral factors, and intervening obstacles, which are illustrated in Lee's model of the migration decision-making process (Figure 4.1). For a person of your age, which factors do you consider would exert the strongest influence on your decision whether or not to move?
2 How do macro-level structural forces influence the migration decisions of individual people?
3 Why does the migration process tend to be quite selective in terms of the age, gender, education and wealth status of those involved?
4 Describe and compare Plates 4.1 and 4.2. Whilst mechanization may help to improve the quality of life and incomes of those who can afford it, do the likely effects of mechanization on levels of agricultural employment, and thus migration, justify the large-scale substitution of people by mechanical farming inputs?
5 How useful are the data contained in Tables 4.3 and 4.4 in helping to explain the incidence of rural–urban migration in Third World countries?

Further reading

Chant, Sylvia (ed) (1992) *Gender and Migration in Developing Countries*, London: Belhaven.
Connell, J. (1976) *Migration from Rural Areas: The Evidence From Village Studies*, Delhi: Oxford University Press.
Fuller, T. D., P. Lightfoot and P. Kamnuansilpa (1985) 'Rural–urban mobility in Thailand: a decision-making approach', *Demography* 22, pp. 565–79.
Lee, Everett S. (1966) 'A theory of migration', *Demography* 3, pp. 47–57.
Nelson, Joan M. (1976) 'Sojourners versus new urbanites: causes and consequences of temporary versus permanent cityward migration in developing countries', *Economic Development and Cultural Change* 24, pp. 721–57.
Rhoda, R. E. (1983) 'Rural development and urban migration: can we keep them down on the farm?', *International Migration Review*, 17.
Riddell, J. B. (1981) 'Beyond the description of spatial pattern: the process of proletarianization as a factor in population migration in West Africa', *Progress in Human Geography* 5, pp. 370–92.
Shrestha, Nanda R. (1987) 'Institutional policies and migration behavior: a selective review', *World Development* 15 (3), pp. 329–45.

Chapter 5

Review questions

1 Where a migrant's absence from his or her home area is timed strictly to coincide with the agricultural cycle, are the effects of the migrant's absence

entirely neutral? Do cash remittances sent from town adequately compensate for the loss of labour and other consequences of out-migration?

2 Describe Figure 5.1. On the basis of your reading of Chapter 5, does the city or the village derive the greater net benefit from migration?

3 To what extent is it possible to isolate the effects of migration from those of other processes of change and development in Third World societies?

4 How do you explain the widespread tendency for migrants' remittances to be used primarily on consumption, as opposed to investment, forms of expenditure?

5 Explain what you understand by the term 'overurbanization', and assess the contribution of rural–urban migration to this phenomenon.

6 Outline some of the 'coping mechanisms' which migrants employ to ease the economic, social and psychological difficulties which may typically be associated with a move to a new location.

Further reading

Catholic Institute for International Relations (1987) *The Labour Trade: Filipino Migrant Workers Around the World*, London: CIIR.

Gilbert, Alan G. and Peter M. Ward (1982) 'Residential movement among the poor: the constraints on housing choice in Latin American cities', *Transactions of the Institute of British Geographers (new series)* 7 (2), pp. 129–49.

Hirabayashi, Lane Ryo (1983) 'On the formation of migrant village associations in Mexico: Mixtec and mountain Zapotecs in Mexico City', *Urban Anthropology* 12 (1), pp. 29–44.

Lipton, Michael (1980) 'Migration from rural areas of poor countries: the impact on rural productivity and income distribution', *World Development*, 8 (1), pp. 1–24.

Plath, J. C., D. W. Holland and J. W. Carvalho (1987) 'Labour migration in Southern Africa and agricultural development: some lessons from Lesotho', *The Journal of Developing Areas*, 21, pp. 159–76.

Rempell, H. and R. A. Lobdell (1978) 'The role of urban to rural remittances in rural development', *Journal of Development Studies*, 14 (3), pp. 324–41.

Rigg, Jonathan (1988) 'Perspectives on migrant labouring and the village economy in developing countries: the Asian experience in a world context', *Progress in Human Geography*, 12 (1), pp. 66–86.

Russell, Sharon Stanton (1986) 'Remittances from international migration: a review in perspective', *World Development* 14 (6), pp. 677–96.

Wood, Charles H. and Terry L. McCoy (1985) 'Migration, remittances and development: a study of Caribbean cane cutters in Florida', *International Migration Review*, 19 (2), pp. 251–77.

Chapter 6

Review questions

1 Which of the four broad policy approaches discussed in Chapter 6 would you expect to be the more effective in tackling the volume and negative effects of rural–urban migration (a) in the short-term, and (b) in the longer-term?
2 By attempting to ease the problems faced by people in urban areas, are not planners also making the cities more attractive to subsequent waves of rural migrants, thereby paving the way to problems of an even greater magnitude in the long-run?
3 Should mobility be an inalienable human right, even when it may lead to considerable hardship and misery for those not directly involved? Do the achievements of the migration programmes in China and Cuba in restricting the pace of urban growth and supporting the development of the countryside suggest that there is a potential role for policies which restrict people's individual rights of movement?
4 'A strong dose of development' offers the best long-term remedy to many of the problems associated with migration in Third World countries. Is migration restricting or strengthening the prospects of development in the Third World?

Further reading

Bilsborrow, R. E., A. S. Oberai and Guy Standing (eds) (1984) *Migration Surveys in Low Income Countries: Guidelines for Survey and Questionnaire Design*, London: Croom Helm.
Oberai, A. S. (ed) (1983) *State Policies and Internal Migration: Studies in Market and Planned Economies*, London: Croom Helm.
Peek, Peter and Guy Standing (eds) (1982) *State Policies and Migration: Studies in Latin America and the Caribbean*, London: Croom Helm.

Index